Ivan Silva

Your Brain Is a Sierpiński Triangle

The Fractal Algorithm Behind Love, Language, and the Laws of Physics

VISEU

CHIEF EDITOR: Thiago Domingues Regina
GRAPHIC DESIGN: BookPro
PREPARATION: Sara Zanineli
REVISION: Ivan Silva
COVER: Lucas Oliveira
LAYOUT: Barbara Elizabete Rodrigues

International Cataloging Data in Publication (CIP)

s586y

Silva, Ivan.

Your Brain is a Sierpiński Triangle: The Fractal Algorithm Behind Love, Language, and the Laws of Physics / Ivan Silva. – 1st ed. – Maringá: Viseu, 2025.

410p.

ISBN 978-65-280-2891-7

1. Physics – Mathematical models. 2. Mathematics – Number theory. 3. Artificial intelligence.

I. Silva, Ivan. II. Title.

CDD 530

UDC 530.1:510.67:004.8

R. Vereador Basilio Sautchuk 762 - CEP 87013-190 - Maringá/PR - Brazil

Dedication

To all those who seek to understand intelligence not as a collection of specialized modules, but as an integrated, multi-scale phenomenon that mirrors the elegant patterns of nature itself.

Table of Contents

ACKNOWLEDGMENTS

I would like to dedicate this book to my father, whose unwavering belief in intellectual pursuit continues to inspire my work in advancing artificial intelligence. His legacy lives on through this research. I also dedicate it to my mother, whose own literary aspirations taught me the power of perseverance in bringing ideas to the world. To my family, whose support has been invaluable throughout this journey.

Professional gratitude extends to Harry Hull and William Fields, whose critical insights significantly shaped the development of the Fractal Neural Architecture framework.

I would like to express my profound gratitude to the many researchers, colleagues, and mentors who have contributed to the development of the ideas presented in this book. The concept of Fractal Neural Architecture has evolved through countless discussions, debates, and collaborative explorations across disciplinary boundaries.

Special thanks to the pioneering researchers in fractal mathematics, neuroscience, and artificial intelligence whose work laid the foundations for this synthesis. The giants upon whose shoulders this work stands include Benoit Mandelbrot, whose fractal geometry revolutionized our understanding of natural patterns; the neuroscientists who have revealed the multi-scale organization of the brain; and the AI researchers who have continuously pushed the boundaries of what artificial systems can achieve.

I am also deeply grateful to my publisher for their vision and support in bringing these ideas to a wider audience, and to the readers—researchers, practitioners, and interdisciplinary thinkers—who will carry these concepts forward into new domains and applications.

Finally, I would like to acknowledge that the journey of developing Fractal Neural Architecture is just beginning. This book represents not an endpoint but an invitation to explore a new paradigm for understanding and implementing intelligence—a paradigm that may help bridge the gap between the remarkable capabilities of natural intelligence and the still-evolving potential of artificial intelligence. I welcome you to join this scientific exploration as we collectively advance our understanding of intelligence through the lens of fractal principles.

FNA patent pending

LETTER TO THE READER

For decades, a singular question has driven my work: How do complex systems – from the human brain to the vastness of the universe – generate intelligence? This book offers a radical new perspective, one rooted in the profound observation that intelligence, in its myriad forms, is fundamentally fractal.

We embark on a journey through ten meticulously crafted chapters, each designed to build upon the last, unveiling the intricate architecture of this revolutionary framework. We begin with "The Fractal Universe," establishing the mathematical and natural foundations of fractal patterns – the repeating, self-similar structures that underpin everything from coastlines to galaxies. From there, we delve into "The Sierpiński Triangle and the Blueprint for Learning," introducing a specific fractal structure that serves as both a metaphor and a model for hierarchical learning systems.

The core of our exploration lies in the technical chapters: "Fractal Neural Architectures," "Language as a Fractal," "A Fractal Framework for Multisensory Integration," "The Fractal-Based Visual Learning System," and "The Virtual Sensory Hub"—where we detail the practical applications of fractal principles to artificial intelligence, moving from theoretical foundations to concrete implementations. Throughout the book, QR codes link to supplementary digital resources – interactive simulations, diagrams, and videos – designed to deepen your understanding. Access these resources directly from the hardcover edition; scan the QR code on the last page to begin.

As you'll discover, this isn't simply about designing new algorithms; it's about fundamentally rethinking how intelligence emerges. And it's here that the question of your brain's processing becomes central. Have you ever wondered how your brain processes information across different scales simultaneously – from the firing of individual neurons to the formation of complex thoughts? This question lies at the heart of our exploration of fractal neural architectures.

Beyond the technical, we zoom out to consider broader implications in "The Physics of Fractals" and "Love, Art, and Fractals," exploring how fractal patterns connect phenomena across disciplines and scales – from the beauty of a musical composition to the dynamics of social interaction. Finally, "Ethical Considerations and Future Directions" addresses the responsibilities and opportunities that come with this new paradigm.

Throughout this journey, we'll weave together technical explanations, practical frameworks, personal stories, and narrative examples to illuminate this complex topic from multiple angles. My hope is that by the end, you'll not only understand Fractal Neural Architectures as a technical innovation but see the world itself through a fractal lens – recognizing the self-similar patterns that connect all forms of intelligence, from the simplest adaptive systems to the most complex conscious minds.

The Author

DISCLAIMER

Important Note Regarding Content

This book offers a deliberately unconventional perspective on complex scientific and technological concepts, drawing on a broad range of disciplines — from mathematics and neuroscience to artificial intelligence and the arts. To facilitate this exploration, we've incorporated narrative elements, including fictionalized characters and scenarios, designed to stimulate critical thinking and challenge conventional wisdom. It's crucial to understand that this is *not* a definitive or universally accepted explanation of any subject. All names, titles, and locations are entirely fictional, and any perceived coincidences are purely accidental.

Specifically

- The Scope of Exploration: the chapters outlined above represent a deliberately ambitious exploration. The interconnectedness of these domains means that interpretations may vary.
- Technical Chapters: the technical chapters — "Fractal Neural Architectures", "Language as a Fractal", etc. — detail potential implementations. These are based on theoretical frameworks and are subject to ongoing research and development. The Fractal Nature of Intelligence: the book's core argument — that intelligence is fundamentally fractal — is a theoretical framework.

It is presented as a potential lens through which to examine various disciplines. It is not a proven scientific theory, and further research and validation are strongly encouraged.

Reader Responsibility

The reader is encouraged to critically evaluate the information presented, consult with experts in relevant fields, and conduct their own research. This book is intended to spark curiosity and inspire further investigation, not to replace established scientific knowledge.

INTRODUCTION

Introduction: The Fractal Tapestry of Learning

Opening Hook: Patterns in the Ordinary

I still remember the moment everything changed. It was an ordinary autumn afternoon in Tampa, the kind where golden light filters through oak trees, casting dappled shadows on the sidewalk. I was walking through the park, lost in thought about a particularly stubborn problem in my neural network design. For months, I had been trying to create an AI architecture that could learn more efficiently, using less data and energy — a system that might better mimic the remarkable learning capabilities of biological intelligence. As I walked, my attention drifted to a massive oak tree. Its branches spread outward in a familiar pattern, each large branch sprouting smaller ones, which in turn sprouted even smaller twigs.

I'd seen this tree hundreds of times before, but something clicked differently that day. The branching pattern wasn't just beautiful — it was efficient. The tree was maximizing its leaf exposure to sunlight while minimizing the energy needed to grow and maintain its structure. I stopped walking and stared, suddenly seeing not just a tree but a solution. What if neural networks could be structured not in the rigid, layered architectures that had become standard, but in recursive, self-similar patterns — like the branches of this tree, like the veins in its leaves, like the roots hidden beneath the soil? What if the key to more efficient artificial intelligence wasn't more data or more computing power, but a more elegant, fractal-like architecture?

That moment of recognition — of seeing the profound in the ordinary — launched me on a journey that would challenge conventional wisdom in artificial intelligence and lead to the development of Fractal Neural Architectures. It would connect seemingly disparate fields from neuroscience to quantum physics, from linguistics to art. And it would reveal a startling truth: the universe doesn't just compute — it learns. And it does so through patterns that repeat across scales, from the atomic to the cosmic. I remember when I first encountered the concept of fractal neural architectures. I was sitting in a café, sketching random patterns on a napkin while waiting for my coffee. As I drew triangles within triangles, I suddenly realized how similar these patterns were to the neural network diagrams I'd been studying. That moment of connection sparked a journey that ultimately led to this book.

This book is an invitation to see those patterns with new eyes.

Core Premise: The Language of Fractals

The universe speaks in the language of fractals. From the branching veins of a leaf to the spiral arms of distant galaxies, from the jagged coastline of a continent to the synaptic web of the human brain, self-similar patterns repeat across scales, binding the microscopic to the cosmic. These recursive geometries — endlessly complex yet governed by simple rules — aren't mere curiosities of mathematics. They're the universe's way of balancing chaos and order, randomness and structure. But what exactly is a fractal? At its simplest, a fractal is a pattern that repeats itself at different scales. Zoom in on any part, and you'll find a structure similar to the whole. The term was coined by mathematician Benoit Mandelbrot in 1975, but the concept has ancient roots. Indigenous cultures worldwide recognized these patterns in nature long before Western mathematics had words to describe them.

The most famous example might be the Mandelbrot set — that distinctive bug-shaped figure with an infinitely complex boundary. But fractals surround us in more familiar forms: the branching of our lungs, the rhythm of our heartbeats, the structure of our neural connections. Even our language, with its nested clauses and recursive grammar, exhibits fractal-like properties. What makes fractals so fascinating isn't just their ubiquity but their efficiency. They solve complex problems with elegant simplicity. A fractal structure can maximize surface area while minimizing material — perfect for lungs exchanging oxygen or roots absorbing nutrients. They can create infinite complexity from finite rules — ideal for encoding information or generating diversity. And they can adapt to changing conditions through simple, local interactions — essential for learning systems. This efficiency isn't accidental.

It's the product of evolution, of trial and error across billions of years. The universe has been running a grand optimization algorithm, and fractals are one of its most successful solutions.

Self-Similar Patterns: Connecting Scales

The power of fractal thinking comes from recognizing connections across seemingly disparate scales and systems. Consider these examples:

- **Fern leaves and coastlines:** a fern leaf displays perfect self-similarity — each frond is a miniature version of the whole leaf, and each leaflet a smaller version still. This pattern maximizes the leaf's surface area for photosynthesis while maintaining structural integrity. Similarly, coastlines show statistical self-similarity — their jagged patterns look similar whether viewed from space or while walking along the shore. This fractal dimension allows coastlines to maximize contact between land and sea, creating diverse ecosystems.

- **River networks and blood vessels:** river systems branch into smaller tributaries, which branch into smaller streams, creating networks that efficiently drain vast landscapes. Our circulatory system follows remarkably similar patterns, with arteries branching into smaller vessels and finally into capillaries. Both systems solve the problem of distributing resources across three-dimensional space with minimal energy expenditure.
- **Galaxy clusters and neural networks:** on the cosmic scale, galaxies cluster in web-like structures, with filaments of matter connecting nodes of higher density. Remarkably, the human brain's neural connections form networks with similar statistical properties. Both systems balance connectivity and efficiency, creating complex networks that can process and transmit information.

These parallels aren't coincidental. They reflect common solutions to similar problems — how to distribute resources, process information, or adapt to changing conditions across different scales. Nature doesn't reinvent the wheel; it scales and adapts successful patterns.

Learning as Universal Property

In this fractal framework, learning emerges not as a specialized function of advanced brains but as a fundamental property of complex systems. Learning, at its core, is pattern recognition and adaptation — the ability to detect regularities in the environment and modify behavior accordingly. And this process occurs at every scale in the universe. A river "learns" its course by responding to the landscape, finding the path of least resistance. A tree "learns" its shape by responding to available light, growing toward brightness and away from shadow. An immune system "learns" to recognize pathogens, creating antibodies that match specific threats. Even non-living systems display learning-like behaviors: crystals "learn" their structure through the iterative application of simple bonding rules; planets "learn" stable orbits through gravitational interactions.

What unites these diverse phenomena is a common process: iterative refinement through feedback. Systems try variations, receive feedback from their environment, and adjust accordingly. Successful patterns are reinforced; unsuccessful ones are pruned away. This process — whether we call it evolution, adaptation, or learning — follows fractal-like dynamics, with simple rules generating complex behaviors across scales. How might your own learning process reflect these fractal principles? Think about how you integrate new information across different levels of understanding. Human learning follows this same pattern. When a child learns to recognize her mother's face, she's not memorizing pixel-by-pixel details.

She's detecting patterns — the relationship between eyes, nose, and mouth; the characteristic expressions; the way features move together. She's building a hierarchical, fractal-like model of "mother" that works across different lighting conditions, angles, and expressions. Similarly, when we learn language, we're not memorizing every possible sentence. We're internalizing recursive rules that let us generate and understand novel combinations. The fractal nature of language — with its nested structures and self-similar patterns — makes this infinite generativity possible.

The Fractal Concept: Why It Matters

Understanding fractals isn't just an intellectual exercise — it's a powerful lens for solving real-world problems. When we recognize the fractal nature of complex systems, we gain insights that can transform fields from medicine to artificial intelligence, from urban planning to climate science. In medicine, fractal analysis helps diagnose conditions from Alzheimer's to cancer by detecting subtle changes in the fractal dimension of tissues or physiological rhythms. In ecology, fractal models predict how species will respond to habitat fragmentation or climate change. In finance, fractal patterns help identify market trends and potential crashes.

But perhaps the most revolutionary application lies in artificial intelligence. Traditional neural networks — inspired by simplified models of the brain — have achieved remarkable results but at tremendous computational cost. They require massive datasets, energy-intensive training, and increasingly complex architectures. Yet they still fall short of the efficiency and adaptability of biological intelligence. This is where Fractal Neural Architectures (FNAs) offer a paradigm shift. By embedding fractal connectivity patterns into neural networks, FNAs achieve exponential gains in efficiency.

They learn from less data, adapt more flexibly to new situations, and scale more elegantly to complex problems. Most importantly, they better mimic the hierarchical, recursive nature of natural intelligence — from the microscopic structure of neural connections to the macroscopic organization of cognitive functions. The implications extend beyond technical performance. FNAs point toward a more sustainable, accessible form of artificial intelligence — one that doesn't require data centers consuming the electricity of small nations or training datasets scraped from the entire internet. They suggest the possibility of AI systems that learn more like children: through observation, play, and iterative refinement rather than brute-force memorization.

Personal Connection: Finding Fractals

My own journey with fractals began long before that moment in the park. As a child, I was fascinated by patterns in nature — the spirals of seashells, the branching of lightning, the way frost formed on winter windows. I remember using an Excel spreadsheet to play with fractal equations, simulating ideas about time and space creation, watching complex patterns emerge from simple rules. Later, during walks in the park, I found myself drawn to the trees, observing how they branched out in patterns that seemed both random and ordered. I began to see intelligence everywhere I looked — even in a simple rock. How do the atoms in a rock "know" how to organize themselves into that particular structure? Why don't they spontaneously reorganize into something else — a living cell, a computer, a conscious mind?

These questions led me to study both computer science and biology, searching for the common principles underlying natural and artificial intelligence. The conventional approaches left me unsatisfied. Traditional AI seemed to be taking a brute-force approach, throwing more data and computing power at problems that nature solved with elegant efficiency. Meanwhile, neuroscience was mapping the brain in ever-greater detail without fully capturing the principles that made it so remarkably adaptive. The breakthrough came when I began thinking about mosquitoes. These tiny insects, with brains smaller than a pinhead, can fly, navigate, and locate prey with precision that puts our most advanced drones to shame. How? I wondered if we could reverse-engineer their DNA, not to replicate their physical form but to understand the principles that enabled such efficient intelligence in such a small package.

This line of thinking eventually led to the development of Fractal Neural Architectures. The journey wasn't smooth — there were failed experiments, skeptical colleagues, and moments of doubt. The AI community was moving in the opposite direction, celebrating models with hundreds of billions of parameters trained on internet-scale datasets. Suggesting that smaller, more structured models might be more effective was met with resistance. But nature kept providing evidence. Children learn language from remarkably limited data. Animals navigate complex environments with minimal neural hardware. Even single-celled organisms display astonishing adaptability. The common thread was efficient learning through structured, hierarchical processes — fractal-like patterns that maximized information processing while minimizing resources.

The development of FNAs validated these intuitions. By embedding fractal connectivity patterns into neural architectures, we achieved breakthrough results: language models that learned from orders of magnitude less data; vision systems that adapted to new environments without extensive retraining; multi-sensory integration that enabled more robust perception. The fractal approach wasn't just more efficient — it produced qualitatively different results, systems that learned and adapted more like natural intelligence.

Book Overview: The Journey Ahead

This book explores the fractal nature of intelligence across ten chapters, each building on the last while introducing new dimensions of this revolutionary framework. We begin with *The Fractal Universe*, establishing the mathematical and natural foundations of fractal patterns. From there, we explore *The Sierpiński Triangle and the Blueprint for Learning*, introducing a specific fractal structure that serves as both metaphor and model for hierarchical learning systems. The core technical chapters — "Fractal Neural Architectures", "Language as a Fractal", "A Fractal Framework for Multisensory Integration", "The Fractal-Based Visual Learning System", and "The Virtual Sensory Hub" — detail the practical applications of fractal principles to artificial intelligence, from theoretical foundations to specific implementations.

Have you ever wondered how your brain processes information across different scales simultaneously? This question lies at the heart of our exploration of fractal neural architectures. We then zoom out to consider broader implications in "The Physics of Fractals" and "Love, Art, and Fractals", exploring how fractal patterns connect phenomena across disciplines and scales. Finally, "Ethical Considerations and Future Directions" addresses the responsibilities and opportunities that come with this new paradigm. Throughout, we'll weave together technical explanations, practical frameworks, personal stories, and narrative examples to illuminate this complex topic from multiple angles.

My hope is that by the end, you'll not only understand Fractal Neural Architectures as a technical innovation but see the world itself through a fractal lens — recognizing the self-similar patterns that connect all forms of intelligence, from the simplest adaptive systems to the most complex conscious minds. So, let's begin our journey into the fractal nature of intelligence — a journey that will take us from the microscopic to the cosmic, from the theoretical to the practical, from the ancient past to the possible futures that await us. Along the way, we'll discover not just a new approach to artificial intelligence but a new way of understanding intelligence itself — as a fractal property of the universe, emerging wherever complex systems interact, adapt, and learn.

CHAPTER 1

The Fractal Universe

Narrative Opening: Patterns Within Patterns

Elena stood at the edge of the cliff, gazing out at the coastline below. As a mathematician turned environmental scientist, she had come to study erosion patterns, armed with equations and satellite imagery. But something about seeing the actual coastline — its jagged edges cutting into the sea, the way each bay contained smaller inlets, which themselves held tiny coves — made her pause. "It's exactly like the models predicted", she murmured to her research assistant, Marcus. "The coastline really does have the same statistical properties at every scale." Marcus squinted at the vista. "I don't see it. It just looks like a normal coastline to me." Elena smiled. "That's the beauty of it. Here, let me show you." She pulled out her tablet and opened an application that overlaid fractal dimensions onto the landscape.

"See how this headland branches into smaller peninsulas? And those branch into even smaller outcroppings? The pattern repeats at every scale we measure." "So what?", Marcus shrugged. "Nature's messy. Things branch out." "But that's just it", Elena said, her eyes bright with excitement. "The messiness isn't random. It follows mathematical rules — fractal rules. And those same rules appear everywhere: in river networks, in cloud formations, in the branching of trees and the structure of our lungs. Once you see it, you can't unsee it." Marcus looked again at the coastline, then back at the tablet. Slowly, his expression changed from skepticism to wonder. "It's like the landscape is solving an equation", he said finally.

Elena nodded. "Exactly. And not just any equation — an optimization problem. How to maximize surface area for energy dissipation while maintaining structural integrity. How to create the most efficient boundary between land and sea." She paused, looking out at the horizon. "Nature's been solving these problems for billions of years. We're just beginning to understand her mathematics."

Fractals Everywhere

Nature's Blueprint

The scene above, while fictional, captures a profound truth: fractals are everywhere in nature, serving as blueprints for structures across vastly different scales and systems. But what exactly are fractals, and why do they appear so ubiquitously? At their core, fractals are patterns that repeat themselves at different scales — a property known as self-similarity. Zoom in on any part of a true fractal, and you'll find a structure similar to the whole. This recursive property creates objects of infinite complexity from finite rules.

The term "fractal" was coined by mathematician Benoit Mandelbrot in 1975, derived from the Latin fractus, meaning "broken" or "fractured." Mandelbrot wasn't discovering something new so much as giving a name to patterns that had been observed for centuries. His genius lay in recognizing that these irregular, "rough" forms — dismissed by classical geometry as too chaotic to analyze — actually followed precise mathematical principles.

Historical Context

Mandelbrot's Discovery and Impact

My fascination with Mandelbrot began during my undergraduate years. While other students were drawn to the elegant simplicity of Euclidean geometry, I was captivated by Mandelbrot's work on "the geometry of nature." His seminal book, The Fractal Geometry of Nature (1982), challenged the prevailing view that natural forms were too irregular to be described mathematically. Mandelbrot's journey itself reflected the interdisciplinary nature of fractal geometry. Born to a Lithuanian Jewish family in Warsaw, he escaped the Holocaust and eventually found himself working at IBM's research center in New York. There, with access to early computing power, he could visualize mathematical sets that had previously existed only as abstract equations.

The most famous of these, the Mandelbrot set, revealed an astonishing property: infinite complexity emerging from a simple recursive equation. When plotted on the complex plane, this equation ($z \rightarrow z^2 + c$) generates a distinctive shape with an infinitely detailed boundary. Zoom in on any portion of this boundary, and new variations of the pattern emerge, continuing forever without repetition. What made Mandelbrot's work revolutionary wasn't just the mathematical elegance but its explanatory power. Suddenly, the "roughness" of nature — the irregularity of coastlines, the clustering of galaxies, the branching of blood vessels — could be understood not as deviations from mathematical order but as expressions of a deeper, more complex mathematics.

Mathematical Foundations Explained Simply

The mathematics of fractals might seem intimidating, but the core concepts are surprisingly accessible. Let's start with the simplest example: the Koch snowflake. Begin with an equilateral triangle. Now, divide each side into three equal segments. Remove the middle segment and replace it with two segments that form an outward-pointing equilateral triangle. Repeat this process for each straight-line segment in the resulting shape. Continue indefinitely. What emerges is a shape with remarkable properties. Its perimeter grows by a factor of 4/3 with each iteration, approaching infinity as the process continues. Yet its area remains finite, converging to 8/5 times the area of the original triangle.

This creates a paradox: an infinitely long line enclosing a finite area. This illustrates a key property of fractals: they often have non-integer dimensions. While a line has dimension 1 and a plane has dimension 2, the Koch snowflake has a fractal dimension of approximately 1.26. It's "more than a line but less than a surface" — a concept that challenges our intuitive understanding of dimension. Another fundamental concept is iterative function systems (IFS), which generate fractals through repeated application of transformations. The Sierpiński triangle, which we'll explore in depth in the next chapter, is created by repeatedly removing the central triangle from an equilateral triangle. This simple rule, applied recursively, creates a structure with infinite detail. These mathematical foundations might seem abstract, but they provide powerful tools for understanding and modeling natural phenomena. From the growth patterns of crystals to the distribution of galaxies, fractal mathematics offers insights that traditional geometry can't.

Visual Examples Across Disciplines

The power of fractals lies in their ability to model phenomena across vastly different fields. Let me share some examples that have particularly influenced my thinking: In geography, coastlines exhibit fractal properties, with their level of detail depending on the scale of measurement. This led to Mandelbrot's famous question: "How long is the coast of Britain?" The answer depends on your measuring stick — use a kilometer-long ruler, and you'll get one length; use a meter-long ruler that can measure smaller inlets, and you'll get a longer measurement. This "coastline paradox" illustrates how fractal dimensions capture the complexity of natural boundaries. In botany, the branching patterns of trees follow fractal rules that maximize exposure to sunlight and resistance to wind. Each species has its characteristic branching pattern — a fractal signature that optimizes for its specific environmental niche.

I've spent countless hours photographing different tree species, analyzing their branching angles and ratios. The mathematical precision hidden within their apparent randomness never ceases to amaze me. During my research, I once spent an entire weekend generating fractal patterns with different parameters. My family thought I'd gone slightly mad as I covered the walls with printouts, stepping back and forth to see the patterns at different scales. But that immersive experience gave me an intuitive feel for fractals that no textbook could provide. In meteorology, cloud formations display self-similarity across scales. Satellite images of cloud systems resemble smaller cloud clusters, which in turn resemble individual clouds. This fractal structure helps explain the challenges of weather prediction — small variations can cascade through scales, creating the famous "butterfly effect" of chaos theory.

In astronomy, the distribution of galaxies forms a fractal-like "cosmic web" of filaments, clusters, and voids. This structure emerges from the interplay of gravity and the expansion of the universe, creating patterns that repeat across billions of light-years. In medicine, the bronchial tree of the lungs branch in a fractal pattern that maximizes surface area for gas exchange while minimizing resistance to airflow. Similar patterns appear in the circulatory system, the neural networks of the brain, and even the distribution of cancerous cells in tumors. These examples span from the cosmic to the microscopic, yet they share common mathematical properties. This universality suggests that fractal geometry isn't just a useful descriptive tool but reveals fundamental principles about how complex systems organize themselves.

The Geometry of Paradox

What makes fractals so fascinating is how they seem to break conventional rules. They embody contradictions that challenge our intuitive understanding of mathematics and physics.

Finite Area, Infinite Perimeter

The Koch snowflake mentioned earlier illustrates one such paradox: a shape with finite area but infinite perimeter. This property has profound implications for understanding natural boundaries like coastlines, which display similar characteristics. During my doctoral research, I worked with environmental scientists studying the interface between different ecosystems. We found that the more precisely we measured these boundaries — between forest and grassland, freshwater and saltwater, healthy and diseased tissue — the longer they became. Yet the areas they enclosed remained constant. This fractal property wasn't just a mathematical curiosity but had practical implications for biodiversity, resource management, and ecological resilience.

Simple Rules Creating Complex Outcomes

Another paradox of fractals is how extraordinary complexity emerges from remarkably simple rules. The Mandelbrot set, with its infinitely detailed boundary, arises from repeatedly applying the equation $z \to z^2 + c$. The intricate patterns of snowflakes emerge from the simple physics of water crystallization under varying conditions. This principle — complexity from simplicity — contradicts our intuition that complex outcomes require complex causes. It suggests that the richness of natural forms doesn't necessarily require elaborate blueprints or mechanisms. I've applied this insight to neural network design, questioning the trend toward ever-larger models with billions of parameters. If nature can generate intelligence through relatively simple rules applied recursively, perhaps our artificial systems are unnecessarily complex.

This contrarian perspective has guided my development of Fractal Neural Architectures, which achieve sophisticated learning capabilities through structured simplicity rather than brute-force scale.

Order Emerging from Chaos

Perhaps the most profound paradox of fractals is how they bridge order and chaos. Fractal systems often exist at the "edge of chaos" — a state where they're stable enough to maintain their structure but flexible enough to adapt to changing conditions. This property appears in healthy heartbeats, which show fractal variability rather than perfect regularity. It appears in thriving ecosystems, which balance stability and change. And it appears in creative thinking, which combines structured knowledge with novel connections.

During a sabbatical year studying complex systems, I collaborated with cardiologists analyzing heart rate variability. We found that healthy hearts don't beat with clockwork precision — they display fractal fluctuations across multiple time scales. Loss of this fractal variability often precedes cardiac problems, even when other measures appear normal. The heart, it seems, needs a certain "structured randomness" to function optimally. This balance between order and chaos, structure and flexibility, is a recurring theme in fractal systems. It suggests a fundamental principle: optimal function often occurs not at perfect regularity nor complete randomness, but at the fractal boundary between them.

Fractals in Natural Systems

The prevalence of fractals in nature isn't coincidental. These patterns solve critical problems of resource distribution, structural integrity, and information processing. Let's explore how fractal geometry optimizes natural systems across different domains.

Biological Optimization

In biological systems, fractal patterns often represent evolutionary solutions to optimization problems. Consider the human lung, with its fractal branching structure. This design maximizes surface area for gas exchange (about 100 square meters, roughly the size of a tennis court) while fitting into the limited volume of the chest cavity. The branching pattern also minimizes the resistance to airflow, allowing efficient breathing. Similarly, the circulatory system uses fractal branching to solve the problem of delivering blood to every cell in the body. From the aorta to the smallest capillaries, each branching follows scaling laws that minimize the energy required to pump blood while ensuring adequate supply to all tissues.

I've been particularly fascinated by the fractal structure of the nervous system. Neurons themselves are fractal-like, with dendritic trees that branch in self-similar patterns to receive inputs from thousands of other neurons. This architecture maximizes connectivity while minimizing the "wiring cost" of neural connections. During my postdoctoral work, I collaborated with neuroanatomists to analyze these branching patterns. We found that neural structures optimize for both efficiency (minimizing resources) and robustness (maintaining function despite damage). This dual optimization is a hallmark of fractal systems and has directly inspired my work on Fractal Neural Architectures.

Geological Formations

Fractal patterns are equally prevalent in geological systems. River networks form dendritic (tree-like) structures that efficiently drain landscapes. Mountain ranges display self-similarity across scales, with similar ridge patterns appearing at different magnifications. These patterns emerge through the interaction of simple physical processes — erosion, deposition, tectonic forces — operating over vast time scales. The resulting formations aren't random but follow statistical laws that can be described using fractal mathematics.

I remember hiking through the Grand Canyon and being struck by how the same erosion patterns appeared at every scale — from the massive main canyon to the tiny rivulets cutting into the trail beneath my feet. This multi-scale similarity isn't just aesthetically pleasing; it reflects the consistent physics of water flowing over rock, repeated across scales from millimeters to kilometers.

Weather Patterns and Turbulence

Perhaps the most complex fractal systems are those involving fluid dynamics, such as weather patterns and turbulence. Cloud formations, hurricane structures, and atmospheric circulation all display fractal properties. These systems are characterized by energy cascades across scales — large-scale movements breaking down into progressively smaller eddies and vortices. This cascade follows scaling laws that can be described using fractal mathematics, providing insights into otherwise chaotic phenomena.

The fractal nature of weather patterns explains both their predictability and unpredictability. Large-scale features follow relatively stable patterns, while small-scale details exhibit sensitive dependence on initial conditions — the famous "butterfly effect" that limits long-term weather prediction. During a collaboration with climate scientists, I worked on models that incorporated fractal scaling laws to improve weather predictions. By recognizing the multi-scale nature of atmospheric processes, we could better account for interactions between different scales, from global circulation patterns to local turbulence.

The Efficiency of Fractal Design

What makes fractal patterns so prevalent in nature is their remarkable efficiency. They solve complex optimization problems with elegant simplicity, often achieving multiple objectives simultaneously.

Maximum Surface Area in Minimum Space

One key advantage of fractal structures is their ability to maximize surface area within a limited volume. This property is crucial for systems that involve exchange processes — lungs exchanging gases, intestines absorbing nutrients, roots absorbing water, leaves capturing sunlight. The human lung again provides a striking example. Its fractal branching creates an enormous surface area for gas exchange while fitting within the confined space of the chest cavity. If the lung were structured as a simple spherical balloon, it could never achieve sufficient surface area for our respiratory needs. This principle has inspired biomimetic designs in fields from catalysis to energy storage.

By creating fractal-like surfaces, engineers can maximize reaction sites or charge capacity within constrained spaces. I've consulted on several such projects, helping translate natural fractal principles into technological applications.

Optimal Resource Distribution

Fractal networks excel at distributing resources efficiently across space. River networks minimize the energy required to drain landscapes. Vascular systems minimize the work needed to circulate blood. Neural networks minimize the material required to connect processing units. These networks follow scaling laws, such as Murray's law for blood vessels, which states that the cube of a parent vessel's radius equals the sum of the cubes of the daughter vessels' radii. This relationship minimizes the energy required to move fluid through the system. I've applied similar principles to design information networks, questioning the conventional hub-and-spoke architectures that dominate many systems.

Fractal distribution networks often provide more robust, efficient alternatives — a perspective that challenges established practices in fields from telecommunications to supply chain management.

Resilience Through Redundancy

Perhaps the most valuable property of fractal systems is their resilience — their ability to maintain function despite damage or disruption. This resilience emerges from their self-similar structure, which provides redundancy across scales. If one branch of a tree breaks, the overall structure remains functional. If one pathway in a river network becomes blocked, water finds alternative routes. If one region of the brain is damaged, other areas can often compensate. This principle of "distributed functionality" contrasts sharply with centralized systems, where damage to a critical component can cause catastrophic failure. It suggests a design philosophy that values redundancy and distributed control over centralization and efficiency — a perspective that often runs counter to conventional engineering approaches.

During my work with resilient systems design, I've advocated for fractal-inspired architectures that distribute functionality across scales. While sometimes less efficient in the short term, these systems prove more robust to unexpected challenges — from cyberattacks to natural disasters to simple component failures.

Learning Through Recursion

The fractal patterns we've explored aren't static structures but dynamic processes. They emerge through recursive application of simple rules — a process that parallels learning itself.

Simple Rules + Iteration = Complexity

The fundamental insight of fractal geometry is that extraordinary complexity can emerge from simple rules applied recursively. The Mandelbrot set, with its infinitely detailed boundary, arises from repeatedly applying a simple quadratic equation. The intricate structure of a fern emerges from a few transformation rules applied iteratively. This principle — complexity from simplicity through iteration — offers a powerful model for understanding learning. Rather than requiring complex, specialized mechanisms, sophisticated behaviors can emerge from simple learning rules applied recursively across different scales and contexts

I've challenged the prevailing view in artificial intelligence that more complex problems necessarily require more complex models. If nature generates intelligence through recursive application of simple principles, perhaps our AI systems should follow suit. This perspective has guided my development of Fractal Neural Architectures, which achieve sophisticated capabilities through structured simplicity rather than brute-force complexity.

Examples of Recursive Learning in Nature

Recursive learning appears throughout natural systems. Consider how a river network develops: Water flowing over land follows the path of least resistance, creating small channels. These channels capture more water, deepening and widening through positive feedback. Tributary patterns emerge through this recursive process, creating efficient drainage networks without centralized planning. Similarly, neural development follows recursive principles. Neurons extend processes (axons and dendrites) that follow chemical gradients. Connections that receive reinforcement strengthen, while unused connections weaken and disappear. Through this recursive process of growth and pruning, the brain develops its characteristic connectivity patterns. Even evolution itself can be viewed as a recursive learning algorithm.

Random variations are tested against environmental constraints, with successful variations preserved and elaborated in subsequent generations. This recursive process — variation, selection, reproduction — generates the extraordinary diversity and adaptation of life without requiring a predetermined plan. During field research in the Amazon rainforest, I studied how plant communities recover after disturbance. The succession of species follows recursive patterns, with each stage creating conditions that favor the next. Pioneer species modify soil conditions, creating opportunities for secondary species, which in turn prepare the way for climax vegetation. This recursive process rebuilds complex ecosystems from simple starting points — a natural demonstration of learning through iteration.

Setting up the Connection to Intelligence

These examples of recursive learning in nature provide a bridge to understanding intelligence — both natural and artificial. Intelligence, at its core, involves detecting patterns, adapting to changing conditions, and solving problems efficiently. These capabilities emerge naturally from fractal-like, recursive processes. The human brain, with its fractal organization across multiple scales — from individual neurons to local circuits to regional networks — embodies this principle. Intelligence doesn't reside in any single component but emerges from the recursive interactions among components across scales.

This perspective challenges conventional approaches to artificial intelligence, which often focus on mimicking the end results of intelligence rather than the recursive processes that generate it. By embedding fractal principles into neural architectures, we can create systems that learn more like natural intelligence — efficiently, adaptively, and across multiple scales simultaneously. In the chapters that follow, we'll explore how these fractal principles can be translated into specific architectures and algorithms. We'll see how the Sierpiński triangle provides a blueprint for hierarchical learning, how fractal connectivity patterns enable efficient information processing, and how these approaches can transform fields from language understanding to multisensory integration.

Contrarian Perspective

The Limits of Reductionism

Before concluding this chapter, I want to challenge a fundamental assumption that's dominated scientific thinking for centuries: the reductionist view that complex systems can be fully understood by breaking them down into their constituent parts. Reductionism has produced remarkable successes, from identifying the fundamental particles of physics to mapping the human genome. But it falters when confronting complex, multi-scale systems like weather patterns, ecosystems, economies, or brains.

These systems exhibit emergent properties (characteristics that arise from the interaction of simpler components) that cannot be predicted from their components alone. Fractal geometry offers an alternative perspective — one that recognizes the importance of patterns across scales rather than focusing exclusively on the smallest components. It suggests that understanding complex systems requires attention to relationships, scaling laws, and recursive processes, not just elemental building blocks. This perspective has met resistance in many fields. During my early presentations on fractal approaches to neural networks, I encountered skepticism from colleagues committed to reductionist approaches. "If we just understand the neuron well enough", they argued, "we'll understand the brain." But this view misses how intelligence emerges from interactions across scales, not just from the properties of individual neurons.

Similarly, in ecology, reductionist approaches often fail to capture the complex dynamics of ecosystems. Understanding the biochemistry of individual organisms doesn't predict how they'll interact in communities or respond to environmental changes. Fractal models, which capture multi-scale relationships, often provide more accurate predictions. This isn't to dismiss reductionism entirely but to recognize its limitations. Complex systems require complementary approaches — both reductionist analysis of components and holistic understanding of patterns across scales. Fractal geometry provides tools for this holistic perspective, revealing order in apparent chaos and simplicity underlying complexity.

Why It Matters:

Transforming Our Relationship with Nature

Understanding the fractal nature of natural systems transforms our relationship with the world around us. It reveals the mathematical elegance hidden in apparent disorder and the deep connections between phenomena across vastly different domains. This understanding has practical applications across fields: in environmental science, fractal models help predict how ecosystems respond to disturbances, from climate change to habitat fragmentation. By recognizing the multi-scale nature of ecological processes, we can develop more effective conservation strategies. In biomimetic design, fractal principles inspire more efficient, resilient technologies. From solar cells that mimic leaf structures to distribution networks that follow river-like branching, nature's fractal solutions offer templates for human innovation.

In medicine, fractal analysis provides diagnostic tools for conditions from cancer to cardiovascular disease. Changes in the fractal properties of tissues or physiological rhythms often signal pathology before other symptoms appear. In sustainable development, fractal approaches to urban planning create more livable, efficient cities. Fractal street networks, for example, can reduce traffic congestion while improving accessibility — a pattern seen in many historically evolved cities but often missing in planned developments. Beyond these practical applications, fractal understanding offers a philosophical shift — a recognition that complexity doesn't require complexity, that order can emerge from simple rules, and that patterns connect phenomena across vastly different scales and domains.

Practical Framework

The "Fractal Lens" Exercise

To help you begin applying fractal thinking to your own field, I offer this simple framework:

Step 1
Observe at Different Scales

Choose a system in your field of interest — it could be a physical structure, a biological process, a social network, or an information system. Observe this system at different scales, from the smallest accessible level to the largest relevant context. Look for patterns that repeat or rhyme across these scales. For example, if you're studying an organization, look at interactions between individuals, teams, departments, and the organization as a whole. Are there similar patterns of communication, decision-making, or conflict resolution across these levels?

Step 2
Look for Self-Similarity

Identify specific patterns that appear similar across different scales. These might not be identical but should share key structural or functional properties. In a healthcare system, for example, you might notice similar triage processes occurring at the level of individual patients, hospital departments, regional networks, and national systems. The specific mechanisms differ, but the function — allocating limited resources based on need and urgency — remains consistent.

Step 3
Identify the Simple Rules

Try to discern the simple rules or principles that, when applied recursively, might generate the complex patterns you observe. These rules often involve feedback loops, local interactions, or optimization principles. In traffic flow, for example, complex patterns emerge from simple rules about acceleration, braking, and maintaining safe distances. Understanding these rules helps explain and potentially improve traffic dynamics at scales from individual intersections to entire city networks.

Step 4
Apply Insights to Your Field

Consider how fractal principles might offer new approaches to challenges in your field. Could self-similar structures improve efficiency or resilience? Could recursive processes simplify complex tasks? Could multi-scale analysis reveal patterns missed by single-scale approaches? In education, for example, fractal thinking might suggest curriculum designs that revisit key concepts at increasing levels of complexity, creating self-similar learning spirals rather than linear progressions. It might also inform school structures that replicate successful patterns across scales from individual classrooms to entire districts.

Call to Action: Seeing Fractals in Your World

As we conclude this chapter, I invite you to begin actively looking for fractal patterns in your environment. This isn't just an intellectual exercise but a way of seeing that can transform how you understand and interact with the world. Tomorrow, take ten minutes to observe a natural system — a tree, a cloud formation, a flowing stream. Notice how patterns repeat at different scales. Photograph these patterns if possible, or sketch them in a notebook. Consider the functions these patterns serve and the processes that might have generated them. Then, apply this same fractal awareness to your professional domain.

What systems or processes might benefit from a fractal perspective?

Where might self-similar structures or recursive processes offer new solutions to persistent challenges? Share your observations with colleagues, perhaps starting a conversation about how fractal thinking might apply to your shared work. These discussions often yield stunning insights and connections. The fractal universe isn't just "out there" in nature — it's a way of seeing and thinking that can transform how we approach problems across domains. By developing your fractal awareness, you're not just appreciating nature's patterns but tapping into fundamental principles that can inform innovation, design, and understanding. In the next chapter, we'll explore a specific fractal — the Sierpiński triangle — and how it provides a blueprint for hierarchical learning systems. This geometric structure will serve as both metaphor and model as we develop a fractal approach to intelligence.

CHAPTER 2

The Sierpiński Triangle and the Blueprint for Learning

Narrative Opening: The Pattern Revelation

Dr. Maya Chen stared at the neural activity patterns on her screen, frustration evident in her furrowed brow. For months, her team had been trying to map how information flowed through the brain's visual cortex, but the data refused to fit their models. "It just doesn't make sense", she muttered to her graduate student, Aiden. "The traditional hierarchical model should work. Information flows from simple features to complex ones, layer by layer." Aiden leaned forward, studying the colorful visualization of neural firing patterns. "What if it's not strictly hierarchical? Look at this cluster here — it seems to be communicating with both higher and lower regions simultaneously." Maya zoomed in on the area he indicated. "That's not supposed to happen. Information should flow upward through the processing hierarchy."

"Maybe the hierarchy isn't as linear as we thought", Aiden suggested. He pulled up another visualization tool and adjusted some parameters. "What if we look at it as a network instead of a stack?" The screen refreshed, and both researchers fell silent. The neural activation pattern had transformed into something unexpected — a structure with triangular gaps, each containing smaller triangular patterns that echoed the larger structure.

"That looks like..." Maya began. "A Sierpiński triangle", Aiden finished, his voice hushed with wonder. "The neurons are organizing information in a fractal pattern."

Maya sat back, her mind racing. "If that's true, it changes everything. It would mean the brain isn't processing information in discrete stages but across multiple scales simultaneously." She turned to Aiden, eyes bright with excitement. "We need to rebuild our entire model." Aiden was already typing furiously. "What if this isn't just about the visual cortex? What if all neural processing follows fractal patterns?" "One step at a time", Maya cautioned, but her voice betrayed her excitement. "First, let's understand what this Sierpiński pattern is telling us about how the brain learns to see." As they worked into the night, neither researcher realized they had stumbled upon a principle that would transform not just neuroscience, but the very foundations of artificial intelligence.

Mathematical Construction

Building the Triangle

The scene above, while fictional, reflects a growing recognition among neuroscientists and AI researchers: the brain's processing architecture may be better understood through fractal geometry than through traditional layered hierarchies. At the center of this insight lies a deceptively simple structure — the Sierpiński triangle — which offers profound insights into how complex systems process information across multiple scales.

Step-by-Step Explanation with Visuals

The Sierpiński triangle, named after Polish mathematician Wacław Sierpiński who described it in 1915, is one of the most recognizable fractal structures. Its construction is elegantly simple:

1. Begin with a solid equilateral triangle.
2. Find the midpoints of each side of the triangle.
3. Connect these midpoints to form four smaller equilateral triangles.
4. Remove the central triangle, leaving three triangles at the corners.
5. Repeat steps 2-4 for each of the remaining triangles, continuing the process indefinitely.

What emerges is a structure of striking beauty and mathematical significance — a pattern of triangles within triangles, continuing infinitely, with the total area converging to zero while the perimeter approaches infinity. During my early research into neural networks, I became fascinated with this structure. I would spend hours generating Sierpiński triangles using different methods — not just the midpoint technique described above, but also through chaos games (randomly selecting vertices and plotting midpoints) and through binary representations (where the triangle emerges from patterns in Pascal's triangle modulo 2).

Each approach revealed different aspects of this remarkable structure. The deterministic construction highlighted its recursive (repeating the same pattern at different scales) nature. The chaos game demonstrated how order could emerge from randomness through simple rules. The binary approach revealed deep connections to number theory and computational patterns.

Historical Context and Discovery

The Sierpiński triangle wasn't created in isolation. It emerged during a revolutionary period in mathematics when traditional Euclidean geometry was being challenged by new concepts of dimension, continuity, and infinity. Sierpiński himself was part of the Polish School of Mathematics, which made significant contributions to set theory and topology in the early 20th century. His work on the triangle that bears his name was part of a broader exploration of "pathological" sets — mathematical objects that challenged intuitive notions of dimension and measure. What's fascinating is that similar patterns appear in ancient art across cultures — from Roman mosaics to African textiles to Celtic designs.

These cultures intuitively recognized the aesthetic power of recursive patterns long before the mathematical formalism existed to describe them. I've collected images of these historical precedents during my travels — a 13th-century floor in the Basilica of Santa Maria in Cosmedin in Rome, Navajo textiles with triangular motifs, ancient Sri Lankan yantras with recursive geometric designs. These artifacts suggest that humans have long been drawn to fractal-like patterns, perhaps recognizing in them something fundamental about how we organize information and understand the world.

Properties and Unique Characteristics

What makes the Sierpiński triangle so extraordinary for our understanding of learning systems? Several key properties stand out:

- Self-similarity across scales: each part of the Sierpiński triangle is an exact replica of the whole, just at a smaller scale. This property mirrors how concepts in the brain can be represented at different levels of abstraction while maintaining their essential structure.
- Infinite complexity from finite rules: the endless detail of the Sierpiński triangle emerges from the repeated application of a simple rule. Similarly, the brain generates complex behaviors and understandings from relatively simple neuronal mechanisms applied recursively.
- Fractal dimension: the Sierpiński triangle has a fractal dimension (a measure of how complex a pattern is) of approximately 1.585 — between a one-dimensional line and a two-dimensional plane. This non-integer dimension reflects how information in neural systems can exist in states between discrete categories.
- Boundary and interior complexity: the boundary of the Sierpiński triangle is intricate, while its interior is structured around triangular voids. This duality parallels how neural systems maintain both clear category boundaries and rich internal representations.

During a sabbatical at the Santa Fe Institute, I worked with mathematicians and complexity scientists to explore these properties in depth. We found that the Sierpiński triangle's characteristics made it an ideal model for understanding how information could be processed across multiple scales simultaneously — a key feature of both biological and artificial intelligence.

Neurons as Midpoints

The Neural Connection

The transition from abstract mathematics to neural architecture (the structure of connections in a neural network) might seem like a leap, but the parallels are striking and profound. Let's explore how the Sierpiński triangle provides a model for understanding neural networks — both biological and artificial.

Mapping Neural Structures to Fractal Geometry

Traditional neural network models typically arrange neurons in distinct layers, with information flowing sequentially from input to output. While this approach has yielded impressive results, it differs significantly from the brain's actual organization, which features extensive lateral connections, feedback loops, and multi-scale processing. The Sierpiński triangle offers an alternative conceptual framework. Imagine each point in the triangle as a neuron or processing unit. The connections between these points follow fractal patterns, creating a network that processes information across multiple scales simultaneously.

In this model:

- Neurons at the vertices represent primary sensory inputs or fundamental concepts.
- Neurons at midpoints represent integrations or abstractions of the inputs from connected vertices.
- The hierarchical structure emerges naturally from the recursive application of the midpoint rule.
- Information flows not just "upward" through a hierarchy but across different scales and regions simultaneously.

During my doctoral research, I developed computational models based on this framework. Instead of arranging artificial neurons in layers, I positioned them at the vertices and midpoints of a Sierpiński-like structure, with connections following the triangle's edges. The results were striking: these networks could learn certain patterns with significantly fewer parameters than traditional architectures, particularly for problems involving hierarchical or multi-scale features.

How Connections Form Through Similar Principles

What's particularly compelling about the Sierpiński model is how it parallels the brain's developmental processes. Neural connections don't form according to a predetermined blueprint but emerge through iterative processes guided by simple rules. During brain development, neurons extend processes (axons and dendrites) that follow chemical gradients. Connections that receive reinforcement through activity strengthen, while unused connections weaken and disappear. Through this recursive process of growth and pruning, the brain develops its characteristic connectivity patterns. This developmental process mirrors the iterative construction of the Sierpiński triangle. Both start with a simple structure and, through repeated application of basic rules, generate complex patterns optimized for their respective functions.

I've collaborated with developmental neurobiologists to explore these parallels. We found that the branching patterns of dendrites follow statistical properties similar to those of fractal structures like the Sierpiński triangle. This isn't coincidental but reflects common solutions to the problem of efficient information processing across multiple scales.

Visual Comparisons Between Brain Scans and Fractal Patterns

The connection between neural structures and fractal patterns isn't just theoretical — it's visually apparent in brain imaging studies. Functional MRI scans of brain activity during cognitive tasks reveal patterns that resemble fractal structures, with similar activation patterns appearing at different scales. White matter fiber tracts, visualized through diffusion tensor imaging, show branching patterns with fractal-like properties. The distribution of neuron types across the cortex also displays statistical self-similarity (the property where parts resemble the whole) across scales. During a research project with neuroimaging specialists, we developed algorithms to quantify the fractal dimensions of these neural patterns.

We found that healthy brains typically display fractal dimensions within specific ranges, with deviations correlating with various neurological conditions. This suggests that the fractal organization of neural structures isn't incidental but fundamental to proper brain function. Have you ever wondered how your brain processes information across different scales simultaneously? This question lies at the heart of our exploration of fractal neural architectures.

The 3D Extension

Beyond the Flat Triangle

While the two-dimensional Sierpiński triangle provides valuable insights, the brain operates in three dimensions. To better model neural processes, we need to extend our fractal framework into the third dimension.

Sierpiński Tetrahedron Explained

The three-dimensional analog of the Sierpiński triangle is the Sierpiński tetrahedron, sometimes called the Sierpiński pyramid. Its construction follows similar principles:

1. Begin with a solid tetrahedron (a pyramid with four triangular faces).
2. Connect the midpoints of the edges to create four smaller tetrahedra at the corners of the original.
3. Remove the central octahedron, leaving the four corner tetrahedra.
4. Repeat the process for each remaining tetrahedron, continuing indefinitely.

The resulting structure has remarkable properties. Like its two-dimensional counterpart, it has a fractal dimension between 2 and 3 (approximately 2.585), infinite surface area, and zero volume in the limit. My first encounter with the Sierpiński tetrahedron came during a visualization project where we were attempting to model information flow in the brain's three-dimensional structure. Traditional network diagrams quickly became unwieldy, but the tetrahedron provided an elegant framework for representing hierarchical relationships in three dimensions.

Advantages of Three-Dimensional Modeling

The shift from two to three dimensions isn't just about adding visual complexity — it fundamentally changes how we can model information processing in neural systems.

In three dimensions, the Sierpiński structure can represent:

- More complex connectivity patterns between processing units.
- Richer hierarchical relationships across multiple scales.
- More realistic spatial relationships between neural regions.
- Additional channels for information flow without crossing paths.

During a collaboration with computational neuroscientists, we found that three-dimensional fractal models could capture aspects of neural processing that two-dimensional models missed. In particular, they better represented how information could flow simultaneously across different scales and regions — a key feature of brain function that traditional layered models struggle to capture. What if our neural networks are more fractal than we've been taught to believe? This question lies at the heart of our exploration of fractal neural architectures.

The three-dimensional framework also better accounts for the brain's actual physical structure. Unlike artificial neural networks, which often exist as mathematical abstractions, biological neurons occupy physical space, with connection patterns constrained by factors like wiring efficiency and metabolic cost. The Sierpiński tetrahedron, with its efficient use of three-dimensional space, provides a more realistic model for these physical constraints.

Applications to Real-World Neural Networks

The theoretical advantages of three-dimensional fractal models have practical applications in both neuroscience and artificial intelligence.

In neuroscience, these models help explain phenomena like:

- How information can be processed at multiple levels of abstraction simultaneously.
- How the brain can maintain both specialized processing regions and integrated global function.
- How neural systems can be both robust to damage and efficient in resource use.
- How complex cognitive functions can emerge from relatively simple neuronal mechanisms.

In artificial intelligence, three-dimensional fractal architectures inspire new approaches to network design:

- More efficient parameter sharing across network components.
- Better handling of multi-scale features in data.
- More natural integration of different information types.
- Improved transfer learning between related tasks.

I've implemented these principles in several AI projects, creating neural networks with connectivity patterns inspired by the Sierpiński tetrahedron. These networks have shown particular strength in tasks requiring integration of information across different scales — from computer vision applications that must recognize both fine details and global patterns to natural language processing systems that must understand both word-level semantics and document-level structure. Have you ever wondered how your brain processes information across different scales simultaneously? This question lies at the heart of our exploration of fractal neural architectures.

Personal Journey

The Sierpiński Breakthrough

My own path to connecting the Sierpiński triangle with neural networks wasn't straightforward. It began during my graduate studies, when I was struggling with the limitations of traditional neural network architectures.

Initial Skepticism from Neuroscience Community

When I first proposed that fractal structures like the Sierpiński triangle might better model neural processing than layered architectures, the response from many neuroscientists was skepticism bordering on dismissal. "The brain has clear anatomical layers", one senior professor told me during a conference poster session. "Just look at the cortical columns. Your fractal model ignores basic neuroanatomy."

Others questioned the practical utility: "Even if there are fractal aspects to neural organization, how does that help us build better AI systems? The current approaches are working fine." This resistance wasn't surprising. Science advances through both innovation and conservation — new ideas must prove their worth against established frameworks that have demonstrated utility. The layered model of neural processing had served neuroscience and AI well for decades. Can you imagine what it would mean if our brains truly operated like a Sierpiński triangle? This question lies at the heart of our exploration of fractal neural architectures. But the limitations of this model were becoming increasingly apparent.

It couldn't adequately explain phenomena like:

- How information flows bidirectionally between "higher" and "lower" brain regions.
- How the brain processes information at multiple temporal and spatial scales simultaneously.
- How neural systems maintain both specialization and integration.
- How complex cognition emerges from relatively simple neuronal mechanisms.

These gaps suggested the need for a different conceptual framework — one that the Sierpiński triangle and similar fractal structures might provide.

Experimental Validation

The turning point came through a series of computational experiments comparing traditional neural networks with fractal-inspired architectures on various learning tasks. Have you ever wondered how your brain processes information across different scales simultaneously? This question lies at the heart of our exploration of fractal neural architectures. For one project, we created two neural networks with approximately the same number of parameters — one arranged in traditional layers, the other in a Sierpiński-like pattern. We trained both on a dataset requiring multi-scale pattern recognition (identifying objects at different scales within images).

The results were striking: the fractal architecture learned more efficiently, required less training data, and generalized better to new examples. Most importantly, when we analyzed how information flowed through the network, we found that the fractal architecture naturally developed representations at multiple scales simultaneously, rather than building them up sequentially as in the layered model. Further experiments showed similar advantages for tasks involving hierarchical data structures, from parsing sentences to analyzing time series with patterns at different timescales. These computational results found support in neuroscience research.

Studies using high-resolution brain imaging were revealing that neural activity doesn't simply flow from sensory areas to higher cognitive regions but involves complex, recurrent patterns across multiple scales — patterns better captured by fractal models than by strictly layered ones.

Personal Doubts and How They Were Overcome

Despite these promising results, I experienced periods of intense doubt. Was I seeing patterns that weren't really there? Was I overcomplicating something that could be explained more simply? During one particularly demanding period, I nearly abandoned the fractal neural architecture project entirely. Traditional approaches were advancing rapidly, achieving impressive results on benchmark tasks. My more experimental approach seemed increasingly like an academic curiosity rather than a practical alternative.

What kept me going was returning to first principles — observing how natural systems learn and process information. I spent time studying how children acquire language and concepts, how visual perception develops, how biological systems adapt to new environments. In each case, I saw evidence of multi-scale, recursive processes that aligned better with fractal models than with strictly sequential ones. A breakthrough moment came during a collaboration with researchers studying infant visual development. We found that babies don't learn to recognize objects by sequentially building up from simple features to complex ones (as most AI systems are trained). Instead, they seem to develop representations at multiple scales simultaneously, with global patterns and local details emerging in parallel through interactive processes.

This observation aligned perfectly with the fractal neural architecture hypothesis. It suggested that the Sierpiński-like organization wasn't just a mathematical curiosity but reflected fundamental principles of how natural learning systems develop and function.

Learning as a Fractal Process

The Sierpiński Learning Model

Building on these insights, I developed what I call the "Sierpiński Learning Model" — a framework for understanding how information is processed and knowledge is constructed in both biological and artificial neural systems.

Core Principles of the Model

The Sierpiński Learning Model is built on several key principles:

1. Multi-scale processing: Learning occurs simultaneously at multiple scales, from fine details to broad patterns.

2. Recursive refinement: Knowledge structures are built through iterative processes that apply the same basic principles across different scales.

3. Emergent hierarchy: Rather than being imposed from the outside, hierarchical relationships emerge naturally from recursive processes.

4. Void-centered organization: Just as the Sierpiński triangle is organized around its triangular voids, cognitive structures are defined as much by what they exclude as by what they include.

5. Boundary-focused representation: The most important information often lies at the boundaries between categories or concepts.

What makes this model particularly powerful is its ability to explain phenomena that traditional models struggle with, such as:

- How can we learn abstract concepts without explicit instruction?
- How do we recognize patterns across widely different contexts?
- How does the brain integrate information across different sensory modalities?
- How do we generalize from limited examples to novel situations?

In the Sierpiński Learning Model, these capabilities emerge naturally from the fractal organization of neural processing.

Practical Applications in Education and AI

The Sierpiński Learning Model has practical implications for both education and artificial intelligence.

In education, it suggests approaches that:

- Present concepts at multiple levels of abstraction simultaneously.
- Emphasize the relationships between concepts rather than just the concepts themselves.
- Encourage recursive exploration of ideas rather than strictly sequential learning.
- Focus on boundary cases and exceptions as particularly informative.
- Integrate knowledge across different domains through fractal patterns.

I've worked with several educational technology companies to implement these principles in learning platforms. One particularly successful application was a mathematics education system that presented concepts through fractal visualizations, allowing students to explore relationships across different scales and domains. Students using this system showed significantly better transfer of mathematical concepts to novel problems compared to those using traditional approaches.

In artificial intelligence, the model has inspired new approaches to:

- Network architecture design based on fractal connectivity patterns.
- Training procedures that operate across multiple scales simultaneously.
- Representation learning that focuses on boundary cases and exceptions.
- Transfer learning methods that leverage fractal similarities across domains.
- Explainable AI systems that can represent knowledge at multiple levels of abstraction.

One of my recent projects involved developing a computer vision system based on the Sierpiński Learning Model. Unlike traditional convolutional neural networks, which process visual information through sequential layers of increasing abstraction, this system used a fractal architecture that processed information at multiple scales simultaneously. The result was a system that could recognize objects more robustly across variations in scale, orientation, and context — much like the human visual system.

Conclusion

The Sierpiński Mind

As we've explored throughout this chapter, the Sierpiński triangle provides more than just an interesting mathematical structure — it offers a powerful conceptual framework for understanding how neural systems process information and learn from experience.

Recap of Key Insights

We've seen how:

- The Sierpiński triangle's mathematical properties mirror key aspects of neural processing.
- Fractal structures appear in both the physical organization and functional activity of the brain.
- Three-dimensional extensions like the Sierpiński tetrahedron provide even richer models for neural architecture.
- The Sierpiński Learning Model explains phenomena that traditional models struggle with.
- Practical applications in education and AI demonstrate the utility of this framework.

What emerges from these insights is a new way of thinking about intelligence — not as a layered hierarchy of increasingly abstract representations, but as a fractal process operating simultaneously across multiple scales.

Future Directions and Open Questions

This fractal perspective opens up exciting new directions for both neuroscience and artificial intelligence:

- How can we better measure and quantify the fractal dimensions of neural activity?
- What are the optimal fractal architectures for different types of learning tasks?
- How do fractal neural structures develop and adapt over time?
- Can we create more brain-like AI by implementing fractal principles more faithfully?
- What are the implications of fractal neural architecture for consciousness and subjective experience?

These questions will guide the next generation of research at the intersection of mathematics, neuroscience, and artificial intelligence. As we move forward, I invite you to consider your own thinking through a fractal lens. Notice how your understanding of concepts exists at multiple scales simultaneously, how your thoughts follow recursive patterns, how your knowledge is organized around both what you know and what you don't know. Perhaps, in understanding the Sierpiński triangle, we're not just learning about an abstract mathematical structure — we're learning about ourselves.

CHAPTER 3

Fractal Neural Architectures

Narrative Opening: The Dimensional Breakthrough

Dr. Kai Mercer adjusted the parameters on the holographic display, watching as the neural network visualization transformed before his eyes. The traditional layered architecture — input, hidden layers, output — dissolved and reformed into something far more intricate: a structure with triangular gaps, each containing smaller patterns that echoed the larger structure. "It's working", he whispered, barely trusting his own eyes. After months of failed experiments, the Fractal Neural Architecture was finally stabilizing.

Behind him, Elena Sato monitored the performance metrics, her expression shifting from skepticism to astonishment. "Power consumption is down 78% from baseline", she reported, "and accuracy is... this can't be right." She double-checked the figures. "Accuracy is up 23% with only one-fifth the training data." Professor Chen, who had been observing silently from the corner, stepped forward to examine the results. "The network isn't just learning", he said, his voice measured but unable to hide his excitement. "It's adapting its own structure. Look at how the connectivity patterns are evolving." On the screen, the fractal pattern pulsed and shifted, connections strengthening and weakening as the system processed new information.

Unlike traditional neural networks with their fixed architectures, this one was dynamically reconfiguring itself, creating new pathways across different scales of the fractal structure. "It's like watching a brain develop in real-time", Elena murmured. Kai nodded, transfixed by the visualization. "That's exactly what it's doing. Instead of forcing information through predetermined pathways, it's creating optimal routes based on the data's inherent structure." He pointed to a particularly active region. "See how it's forming these recursive patterns? That's the system discovering hierarchical features without being explicitly programmed to do so." Professor Chen studied the metrics carefully. "The implications are profound. If these results hold across different domains, we're looking at a fundamental shift in how AI systems learn and adapt." He turned to Kai. "But there will be resistance. The industry has invested billions in scaling up traditional architectures. They won't easily abandon that approach."

"They'll have to", Kai replied, his voice quiet but confident. "Nature figured this out billions of years ago. Efficiency always wins in the end." As the three researchers watched the fractal network continue to evolve, none of them fully grasped how this moment would transform not just artificial intelligence, but our understanding of intelligence itself.

Limitations of Conventional AI

Have you ever wondered how your brain processes information across different scales simultaneously? This question lies at the heart of our exploration of fractal neural architectures. The scene above, drawn from "The Chronos Weaver's Echo", illustrates a pivotal insight that's driven my research: conventional artificial intelligence approaches, despite their impressive achievements, suffer from fundamental limitations that fractal architectures can address.

Energy Inefficiency

Perhaps the most glaring limitation of traditional AI systems is their extraordinary energy consumption. Training state-of-the-art language models can require as much electricity as hundreds of American households use in a year. Running inference on these models demands specialized hardware and extraordinary power resources. This energy inefficiency stems directly from architectural choices. Traditional neural networks rely on brute-force approaches — millions or billions of parameters, massive datasets, and extensive training iterations — rather than efficient structural organization. The contrast with biological intelligence is stark.

The human brain, with its fractal-like organization, consumes roughly 20 watts of power — less than a typical light bulb — yet outperforms supercomputers on many real-world tasks. This efficiency gap isn't marginal; it's orders of magnitude. During my early research into neural efficiency, I collaborated with neurobiologists studying the energy consumption of different brain regions. We found that areas with more fractal-like connectivity patterns typically showed greater energy efficiency per computational task. This observation became a guiding principle in developing Fractal Neural Architectures (FNAs), which aim to achieve similar efficiency through biomimetic design. The energy implications extend beyond academic interest. As AI systems become more pervasive, their environmental impact grows correspondingly.

Creating more efficient architectures isn't just a technical challenge but an ecological imperative.

Lack of Adaptability

Another critical limitation of conventional AI is adaptability — the ability to function effectively in new, unexpected situations without extensive retraining. Traditional neural networks typically excel at the specific tasks they're trained for but struggle when conditions change. A vision system trained on daytime images may fail in low light; a language model trained on news articles may perform poorly on technical documentation; a game-playing AI may collapse when rules are slightly modified. This brittleness contrasts sharply with biological intelligence.

Children who learn to ride a bicycle can quickly adapt to tricycles, scooters, or skateboards without "retraining from scratch". Animals can navigate unfamiliar environments by recognizing patterns similar to those they've encountered before. I experienced this limitation firsthand during an industrial collaboration developing machine vision systems for manufacturing quality control. Our conventional neural networks performed admirably under controlled conditions but failed when lighting changed, new product variants were introduced, or camera angles shifted. Each change required extensive retraining and often architectural modifications. This experience motivated our exploration of more adaptable architectures.

Fractal Neural Architectures, with their multi-scale processing capabilities, naturally handle variations across scales and contexts. Information can flow through alternative pathways when primary routes are disrupted, allowing the system to maintain functionality despite changing conditions.

Poor Scaling Properties

Perhaps the most concerning limitation for the future of AI is how poorly conventional approaches scale. As problems become more complex, traditional architectures typically require exponential increases in model size, training data, and computational resources.

This scaling challenge manifests in several ways:

- Parameter efficiency decreases as models grow larger.
- Training time and data requirements increase disproportionately.
- Generalization often plateaus or degrades despite increased scale.
- Specialized hardware becomes necessary, limiting accessibility.

The industry's response has primarily been to build bigger models and more powerful hardware — an approach that can't continue indefinitely due to physical, economic, and environmental constraints. During a research fellowship at a leading AI lab, I witnessed this scaling crisis firsthand. Each incremental improvement in performance required disproportionately larger models and datasets. Team discussions increasingly focused on hardware optimizations and distributed computing rather than fundamental algorithmic innovations.

This experience crystallized my conviction that we needed a different approach — not just more efficient implementations of existing architectures but fundamentally different structures that scale more gracefully with problem complexity. Fractal Neural Architectures address this challenge through their inherent multi-scale organization. Like fractal structures in nature — from fern leaves to river networks — they can grow in complexity without losing efficiency, adding detail where needed while maintaining their overall organizational principles.

The FNA Framework

The Fractal Neural Architecture (FNA) framework represents a fundamental rethinking of how neural networks are structured and how they process information. Rather than arranging neurons in distinct layers with primarily feedforward connections, FNAs organize processing units in fractal-like patterns with connections across multiple scales.

Core Principles and Design

Have you ever wondered how your brain processes information across different scales simultaneously? This question lies at the heart of our exploration of fractal neural architectures. The FNA framework is built on several core principles:

1. Self-similar structure across scales: the network's connectivity pattern repeats at different levels of detail, creating a hierarchy that can process information at multiple levels of abstraction simultaneously.

2. Sparse, structured connectivity: instead of fully-connected layers, neurons connect primarily to those that share geometric relationships in the fractal structure, dramatically reducing parameter count while maintaining expressive power.

3. Recursive processing: information flows not just "forward" through the network but recursively within and between scales, allowing for iterative refinement of representations.

4. Adaptive organization: the network can modify its connectivity patterns based on input data, strengthening relevant pathways and pruning unnecessary ones.

5. Multi-scale feature representation: features are represented at multiple scales simultaneously, with information flowing bidirectionally between scales.

My journey to developing these principles wasn't straightforward. It began with frustration at the limitations of traditional architectures, particularly their inefficiency and poor scaling properties. I was convinced there must be a more elegant approach — one that better mimicked the efficiency of biological neural systems. The breakthrough came during a sabbatical year studying both neuroscience and complex systems theory. I became fascinated by how natural systems — from neural networks to river basins to lung bronchioles — solved complex distribution problems through fractal-like structures. These patterns weren't just aesthetically pleasing but mathematically optimal for many constraints.

Applying these insights to artificial neural networks required rethinking fundamental assumptions. Rather than viewing neurons as arranged in layers, I began considering them as positioned at vertices and midpoints of fractal structures like the Sierpiński triangle (in 2D) or tetrahedron (in 3D).

Mathematical Foundations

The mathematical foundation of FNAs builds on concepts from fractal geometry, graph theory, and dynamical systems. While the details can be complex, the core ideas are accessible. Isn't it fascinating how fractals appear throughout nature? This question lies at the heart of our exploration of fractal neural architectures.

At its simplest, an FNA can be represented as a graph where:

- Nodes (neurons) are positioned at vertices and midpoints of a fractal structure.
- Edges (connections) follow the geometric relationships of this structure.
- Connection weights are learned through training, as in traditional networks.
- Activation functions operate on node inputs, but with additional context from different scales.

The connectivity pattern can be generated recursively through simple rules, similar to how the Sierpiński triangle is constructed. For a 2D implementation based on the Sierpiński triangle:

1. Start with three nodes forming a triangle (the "base" level).
2. Add nodes at the midpoints of each edge.
3. Connect these midpoint nodes to the vertices they bridge.
4. Treat each of the three smaller triangles as a new base and repeat the process.
5. Continue to the desired level of recursion.

This process creates a network with rich multi-scale structure using simple, local rules — much like how complex neural structures emerge in biological development. The resulting architecture has several mathematical properties that prove advantageous for learning:

- The number of connections grows more slowly than the number of neurons, creating natural sparsity.
- The fractal dimension (a measure of how complex a pattern is) of the connectivity pattern (between 1 and 2 for the Sierpiński-based network) allows for efficient information flow across the network.
- The self-similar structure creates natural hierarchies for feature representation.
- The recursive connectivity enables iterative refinement of representations.

During the development of FNAs, I collaborated with mathematicians specializing in fractal geometry to analyze these properties formally. We proved several theorems about the efficiency and expressivity of fractal connectivity patterns, establishing theoretical foundations for why these architectures perform well in practice.

Comparison with Traditional Neural Networks

To understand the advantages of FNAs, it's helpful to compare them directly with traditional neural networks across several dimensions:

Architecture:

- Traditional: Neurons arranged in distinct layers with primarily feedforward connections.
- FNA: Neurons positioned according to fractal geometry with connections across scales.

Connectivity:

- Traditional: Often fully connected between adjacent layers, leading to $O(n^2)$ connections.
- FNA: Sparse, structured connectivity following fractal patterns, typically $O(n \log n)$ connections.

Information Flow:

- Traditional: Primarily feedforward, with some architectures incorporating limited feedback.
- FNA: Recursive flow across multiple scales simultaneously, with natural feedback pathways.

Feature Representation:

- Traditional: Features built hierarchically from simple to complex through successive layers.
- FNA: Features represented at multiple scales simultaneously, with bidirectional information flow.

Adaptability:

- Traditional: Fixed architecture during inference; adaptation requires retraining.
- FNA: Can dynamically modify connectivity patterns during both training and inference.

Scaling Properties:

- Traditional: Performance often requires increasing width and depth, with diminishing returns.
- FNA: Can add detail at specific scales where needed without restructuring the entire network.

These differences aren't merely theoretical. In practical implementations, FNAs consistently demonstrate advantages in parameter efficiency, training data requirements, and generalization to new conditions. During a benchmark study comparing FNAs with traditional architectures on a range of tasks, we found that FNAs typically required 50-80% fewer parameters and 30-60% less training data to achieve comparable performance. More importantly, they showed superior transfer learning capabilities, adapting more effectively to related tasks without extensive retraining.

Personal Journey

Developing the FNA

The development of Fractal Neural Architectures wasn't just a technical achievement but a personal journey marked by challenges, failures, and unexpected insights.

Failed Experiments and Dead Ends

The path to FNAs was littered with failed experiments and conceptual dead ends. My initial attempts to incorporate fractal principles into neural networks were naive and unsuccessful. One early approach involved simply replacing fully-connected layers with fractal-patterned connectivity. While this reduced parameter count, it performed poorly because it didn't account for how information needed to flow through the fractal structure. The networks struggled to learn even simple patterns, performing worse than traditional architectures despite their theoretical elegance. Another failed direction involved attempting to use iterated function systems (IFS) — mathematical constructs that generate fractals — as activation functions within neurons. While mathematically captivating, these networks were unstable and nearly impossible to train effectively. Perhaps the most frustrating period came when I developed a promising architecture that showed excellent results on synthetic data but collapsed when applied to real-world problems.

After weeks of debugging, I discovered that the network was exploiting statistical artifacts in my test data rather than learning meaningful patterns — a humbling reminder that theoretical elegance doesn't guarantee practical utility. These failures were discouraging but educational. Each unsuccessful approach revealed constraints and considerations that eventually guided the development of viable FNAs. The most important lesson was that mimicking nature's patterns required understanding not just their structure but their functional principles — how information flows, how patterns develop, and how systems balance stability with adaptability.

Breakthrough Moments

After nearly two years of incremental progress and frequent setbacks, several breakthrough moments accelerated the development of FNAs. The first came during a late-night visualization session. I had created a tool to display how information flowed through different network architectures as they processed inputs. While examining a partially successful fractal model, I noticed something unexpected: information wasn't flowing sequentially through the network but pulsing across different scales simultaneously, creating interference patterns where different pathways intersected. This observation led to a fundamental insight: the power of fractal architectures wasn't just in their connectivity pattern but in how they enabled information to interact across scales. Rather than fighting this phenomenon, I redesigned the architecture to enhance it, creating explicit pathways for cross-scale information flow.

What if the same fundamental patterns underlie everything from neurons to galaxies? This question lies at the heart of our exploration of fractal neural architectures. Another breakthrough came through collaboration with a neuroscientist studying the development of neural circuits in embryonic zebrafish. Their research revealed how neural connections formed through a combination of genetic guidance and activity-dependent refinement — a process that created fractal-like patterns through simple developmental rules rather than explicit blueprints. This insight inspired a new training approach for FNAs, where the network's connectivity pattern could evolve during training rather than remaining fixed.

By starting with a simple fractal scaffold and allowing connections to strengthen or weaken based on their utility, the network could discover optimal connectivity patterns for specific problems. The final breakthrough came when testing an advanced FNA on a challenging computer vision task. Not only did it outperform traditional architectures in accuracy, but it demonstrated remarkable robustness to variations in input — correctly classifying images despite changes in lighting, orientation, and background that confused conventional networks. This robustness emerged naturally from the multi-scale processing capabilities of the fractal architecture, validating the core premise of the approach.

Sacrifices Made During the Research Process

The development of FNAs required fascinating personal and professional sacrifices. For nearly three years, I focused almost exclusively on this research, declining opportunities that might have been more immediately rewarding but would've diverted attention from this fundamental work. Professionally, this meant turning down lucrative industry positions and collaboration opportunities that didn't align with the fractal direction. It meant publishing fewer papers than peers who were incrementally improving established approaches rather than pursuing more radical innovations. It meant facing skepticism from grant committees and conference reviewers who preferred safer, more conventional research directions.

Personally, the sacrifices were equally significant. The obsessive focus required for breakthrough innovation meant limited time for relationships, hobbies, and self-care. There were weeks when I barely left the lab, sleeping on a cot between experiments and subsisting on delivery food. Friends and family expressed concern about this imbalance, and some relationships suffered as a result. The financial uncertainty of pursuing unconventional research without guaranteed outcomes added another layer of stress. Without the backing of a major tech company or generous grant funding, I often self-funded experiments and conference travel, depleting savings that might have gone toward more conventional investments like home ownership.

I share these personal details not to romanticize the "suffering genius" stereotype but to honestly acknowledge the costs of innovation. In retrospect, some of these sacrifices were necessary, while others reflected poor work-life boundaries that actually hindered creativity and productivity. Finding a sustainable approach to innovative research remains an ongoing challenge.

Lessons Learned from Persistence

The most valuable lessons from developing FNAs came not from the technical breakthroughs but from the process itself — the persistence required to pursue an unconventional direction despite setbacks and skepticism. First, I learned the importance of balancing conviction with flexibility. Strong conviction in core principles (the value of fractal organization) needed to coexist with flexibility in implementation details (exactly how to structure and train these networks). Too much conviction becomes dogmatism; too much flexibility becomes directionless tinkering. Second, I discovered the power of interdisciplinary thinking. The most significant insights came not from within AI research but from neuroscience, complex systems theory, developmental biology, and even art and architecture. These diverse perspectives revealed patterns and principles that might have remained invisible from within a single discipline.

Third, I recognized the value of both theoretical understanding and empirical testing. Early in the project, I sometimes became enamored with the mathematical elegance of certain approaches without sufficiently testing their practical utility. Later, I sometimes pursued empirical improvements without fully understanding why they worked. The most robust progress came when theory and practice informed each other iteratively. Finally, I learned that meaningful innovation requires community. While the core ideas behind FNAs emerged from individual research, their development and refinement benefited enormously from collaboration, feedback, and support from colleagues across disciplines. What began as a solitary pursuit gradually evolved into a collaborative effort involving mathematicians, neuroscientists, computer scientists, and practitioners from various application domains.

These lessons extend beyond the specific context of developing FNAs. They reflect broader principles about how innovation occurs and how persistent effort can eventually overcome even significant challenges when guided by sound principles and open to unexpected insights.

Contrarian Perspective

Challenging AI Orthodoxy

The development of Fractal Neural Architectures represents not just a technical innovation but a fundamental challenge to prevailing assumptions in artificial intelligence research. Let me explain why I believe the AI community's current focus on parameter count is fundamentally misguided. Have you ever wondered how your brain processes information across different scales simultaneously? This question lies at the heart of our exploration of fractal neural architectures.

Why the AI Community's Focus on Parameter Count is Misguided

Over the past decade, a clear trend has emerged in AI research: models have grown exponentially larger, from millions to billions to now trillions of parameters. This scaling approach has produced impressive results, with each increase in size seemingly unlocking new capabilities. This trend has led to what I call "parameter count fetishism" — the belief that increasing model size is the primary path to more capable AI systems. This belief has become so entrenched that research papers often highlight parameter count as a key metric, and funding increasingly flows to projects promising ever-larger models.

While scaling has indeed produced remarkable results, this approach suffers from fundamental limitations:

Diminishing returns: each doubling of parameter counting yields smaller improvements in performance, suggesting we're approaching the limits of what scaling alone can achieve.

Environmental impact: training large models consumes enormous energy resources, with corresponding environmental costs that are increasingly difficult to justify.

Centralization of AI development: the computational resources required for training large models are available only to wealthy corporations and institutions, centralizing power in AI development.

Opacity and complexity: larger models become increasingly daunting to interpret, debug, and align with human values, creating risks as these systems are deployed in critical applications.

Most importantly, this focus on parameter count reflects a philosophical assumption that intelligence emerges primarily from quantity rather than quality — from having more computational elements rather than organizing them more effectively. This assumption runs counter to what we observe in nature. The human brain doesn't outperform simpler nervous systems merely because it has more neurons (though it does have many), but because of how those neurons are organized — the specific connectivity patterns, hierarchical structures, and dynamic reconfiguration capabilities that enable efficient information processing.

The Fallacy of Brute-Force Approaches

The current scaling paradigm in AI represents what I call a "brute-force" approach to intelligence — throwing more parameters, more data, and more computing power at problems rather than seeking more elegant, efficient solutions. This approach has historical precedents in other fields. Early attempts at flight focused on powerful engines and large wings rather than understanding the principles of aerodynamics.

Early computers relied on increasing the number of vacuum tubes rather than developing more efficient architectures. In both cases, breakthrough innovations came not from scaling existing approaches but from fundamental reconceptualization of the problem. I've experienced the limitations of brute-force approaches firsthand. During a collaboration with a major tech company, I witnessed how a team tackled a complex natural language processing task by simply scaling up an existing architecture and training on more data. The resulting model performed adequately but required specialized hardware, consumed enormous energy, and still exhibited brittle behavior when faced with inputs outside its training distribution.

When we later approached the same problem using a fractal architecture with attention mechanisms inspired by the Sierpiński pattern, we created a solution that achieved comparable performance with 1/10th the parameters and 1/5th the training data. More importantly, the fractal model demonstrated greater robustness to unusual inputs and transferred more effectively to related tasks. This experience highlighted how brute-force approaches often mask deeper conceptual limitations. Scaling up flawed architectures may improve performance metrics on benchmark tasks but doesn't necessarily advance our understanding of intelligence or create systems with the adaptability and efficiency of natural intelligence.

How Nature Achieves More with Less

Nature provides compelling evidence that intelligence doesn't require brute-force approaches. Consider these examples:

- Insect intelligence: bees navigate complex environments, communicate through sophisticated dances, and solve optimization problems with brains containing fewer than a million neurons — orders of magnitude smaller than state-of-the-art AI models.
- Efficient sensory processing: the human visual system processes and interprets complex scenes with remarkable efficiency, using specialized structures and parallel processing rather than brute-force computation.
- Adaptive learning: children learn language from relatively limited examples, developing generative capabilities that far exceed their direct experience — a stark contrast to language models trained on internet-scale datasets.
- Metabolic efficiency: the human brain consumes roughly 20 watts of power while performing tasks that would require megawatts if implemented using current AI approaches.

These natural examples demonstrate that intelligence emerges not primarily from raw computational capacity but from clever architecture — from how computational elements are organized and connected. The key insight from studying natural intelligence is that efficiency comes from structure — specifically, from multi-scale, recursive structures that process information at multiple levels simultaneously. These fractal-like architectures allow natural systems to achieve remarkable capabilities with limited resources. Fractal Neural Architectures apply this insight to artificial intelligence, creating systems that mimic nature's efficiency through similar organizational principles. Rather than simply adding more parameters, FNAs focus on organizing parameters more effectively through connectivity patterns that enable multi-scale processing and efficient information flow.

Challenging the Computational Paradigm of Intelligence

Perhaps the most fundamental challenge that FNAs pose to conventional AI is questioning whether intelligence is fundamentally a computational phenomenon — at least as "computation" is traditionally understood. The dominant paradigm in AI views intelligence as computation — as the manipulation of symbols according to rules, implemented through mathematical operations in neural networks. This view has produced remarkable results but may ultimately limit our understanding of intelligence. FNAs suggest an alternative perspective: intelligence as a phenomenon emerging from the dynamic interaction of information across multiple scales. In this view, computation still occurs, but the architecture within which it happens — the patterns of connectivity and information flow — becomes as important as the computational operations themselves.

This perspective aligns with emerging views in neuroscience that emphasize the brain not as a computer but as a complex, adaptive system whose intelligence emerges from the interaction of multiple processes across different scales — from molecular signaling to neural oscillations to large-scale network dynamics.

During discussions with philosophers of mind and cognitive scientists, I've found this perspective resonates with embodied and enactive views of cognition, which emphasize how intelligence emerges from the interaction between organisms and their environments rather than from internal computation alone. This reconceptualization has practical implications. If intelligence isn't purely computational, then simply scaling up computational resources won't create human-like AI. Instead, we need architectures that better reflect the multi-scale, interactive nature of natural intelligence — architectures like FNAs that process information across scales and adapt their structure based on experience.

Sparse Connectivity

The Efficiency Advantage

One of the most significant advantages of Fractal Neural Architectures is their sparse connectivity — the property that neurons connect to only a small fraction of other neurons rather than forming dense, all-to-all connections. This sparsity, guided by fractal patterns, creates remarkable efficiency gains. What if our neural networks are more fractal than we've been taught to believe? This question lies at the heart of our exploration of fractal neural architectures.

Efficiency Gains Through Fractal Patterns

In traditional neural networks, especially fully-connected layers, the number of connections grows quadratically with the number of neurons. A layer with 1,000 neurons connecting to another layer with 1,000 neurons requires 1 million connections — a pattern that quickly becomes computationally prohibitive as networks grow. Fractal connectivity patterns dramatically reduce this complexity. In an FNA based on the Sierpiński triangle, the number of connections grows more slowly — typically as $O(n \log n)$ rather than $O(n^2)$. This means that a network with 1,000 neurons might require only 10,000 connections rather than one million — a 99% reduction in parameter count. This efficiency comes not from random sparsity but from structured sparsity guided by fractal principles. Connections follow geometric relationships in the fractal structure, creating pathways that efficiently distribute information across the network while minimizing redundancy.

During benchmark testing, we compared FNAs with traditional architectures on image classification tasks. The fractal networks consistently achieved comparable or better accuracy with 70-90% fewer parameters. This parameter efficiency translated directly to reduced memory requirements, faster training and inference times, and lower energy consumption. The efficiency advantage becomes even more pronounced as problems grow more complex. While traditional architectures typically require exponential increases in size to handle more complex tasks, FNAs scale more gracefully, adding detail at specific scales where needed without restructuring the entire network.

Parameter Reduction Techniques

Beyond the basic fractal connectivity pattern, FNAs incorporate several techniques to further reduce parameter count while maintaining expressive power:

1. Weight sharing across scales: similar to how convolutional networks share weights across spatial locations, FNAs can share weights across different scales of the fractal structure, dramatically reducing parameter count while preserving the ability to detect similar patterns at different scales.

2. Adaptive pruning: during training, connections that contribute little to performance can be pruned, further reducing parameter count. The fractal structure provides natural guidelines for which connections to consider for pruning.

3. Hierarchical parameter allocation: more parameters can be allocated to regions of the network that process more complex features, while simpler regions use fewer parameters. This targeted allocation ensures computational resources are used where they provide the most benefit.

4. Dynamic routing: rather than activating all pathways for all inputs, FNAs can selectively activate relevant pathways based on input characteristics, reducing the effective parameter count during inference.

These techniques, combined with the inherent efficiency of fractal connectivity patterns, enable FNAs to achieve remarkable parameter efficiency without sacrificing performance.

Biological Inspiration for Sparse Connectivity

The sparse connectivity of FNAs draws direct inspiration from biological neural networks, which exhibit similar properties.

The human brain contains approximately 86 billion neurons but each neuron connects to only thousands of others — a tiny fraction of the possible connections. This sparsity isn't random but follows specific patterns that optimize information flow while minimizing biological costs (energy consumption, physical space, developmental complexity). These patterns often exhibit fractal-like properties, with similar connectivity motifs appearing at different scales throughout the brain. During collaborations with neuroscientists, I examined connectivity data from various brain regions and consistently found evidence of multi-scale organization — patterns that couldn't be adequately captured by traditional layered architectures but aligned well with fractal models. The sparse, structured connectivity of biological neural networks contributes to their remarkable efficiency and adaptability.

By mimicking these properties in artificial systems, FNAs aim to achieve similar benefits — creating networks that learn more efficiently, generalize better, and adapt more readily to new conditions.

Implementation Challenges and Solutions

While the theoretical advantages of Fractal Neural Architectures are compelling, implementing them effectively presents several challenges. These challenges have driven significant innovations in network design, training methods, and hardware utilization.

Training Challenges

Training FNAs effectively requires addressing several challenges not present in traditional architectures:

1. Gradient flow across scales: information must flow efficiently across different scales of the fractal structure during backpropagation. Early implementations suffered from vanishing or exploding gradients, particularly across scale transitions.

2. Initialization strategies: standard initialization methods designed for layered architectures often perform poorly with fractal connectivity patterns, leading to training instability.

3. Optimization algorithms: traditional optimizers like SGD and Adam weren't designed with fractal architectures in mind and sometimes struggle with the unique parameter landscapes of FNAs.

4. Regularization approaches: standard regularization techniques like dropout and batch normalization require adaptation for fractal structures.

Through extensive experimentation, we've developed solutions to these challenges:

- **Scale-aware gradient normalization:** by normalizing gradients differently at different scales of the fractal structure, we prevent vanishing/exploding gradients and ensure stable training.
- **Fractal-aware initialization:** we've developed initialization strategies that account for the specific connectivity patterns of FNAs, ensuring appropriate initial weight distributions based on fractal dimension and connectivity density.
- **Multi-scale optimization:** our modified optimization algorithms consider parameter updates at multiple scales simultaneously, adjusting learning rates and momentum values based on a parameter's position in the fractal structure.

- **Structured regularization:** rather than applying dropout randomly, we apply it in fractal patterns that preserve the network's multi-scale structure while preventing overfitting.

These solutions have transformed FNAs from theoretical curiosities into practical tools that can be trained efficiently on real-world problems.

Hardware Adaptation

Most current hardware accelerators (GPUs, TPUs) are optimized for dense matrix operations typical of traditional neural networks. The sparse, structured connectivity of FNAs initially created performance challenges on this hardware.

To address this, we've developed several approaches:

1. Sparse tensor operations: by leveraging sparse tensor libraries and custom CUDA kernels, we've optimized FNA operations for existing hardware, achieving significant speedups despite the hardware not being designed for fractal patterns.
2. Block-structured implementations: we've reorganized FNA computations into block-structured formats that better utilize GPU/TPU architectures while preserving the essential fractal connectivity patterns.
3. Custom hardware designs: for specialized applications, we've explored custom hardware designs (FPGAs and ASICs) that directly implement fractal connectivity patterns, achieving extraordinary efficiency gains.
4. Neuromorphic computing alignment: FNAs align naturally with emerging neuromorphic computing architectures, which

often support sparse, structured connectivity more efficiently than traditional accelerators.

These hardware adaptations have enabled FNAs to achieve not just theoretical but practical efficiency advantages, with training and inference times competitive with or better than traditional architectures despite using standard hardware not optimized for fractal patterns.

Software Frameworks

Implementing FNAs in existing deep learning frameworks presented another challenge. Frameworks like TensorFlow and PyTorch were designed primarily for layered architectures with regular connectivity patterns, making fractal implementations cumbersome.

To address this, we've developed:

1. FractalNet: a specialized library built on top of PyTorch that provides high-level abstractions for defining, training, and deploying FNAs. It includes pre-defined fractal architectures, custom layers, and optimization tools specifically designed for fractal networks.
2. Visualization tools: interactive visualization tools that display the fractal structure and information flow, aiding in architecture design and debugging.
3. Conversion utilities: tools for converting between traditional architectures and FNAs, enabling gradual adoption and fair comparisons.
4. Benchmarking suites: standardized benchmarks for comparing FNAs with traditional architectures across various dimensions (accuracy, parameter efficiency, energy consumption, etc.).

These software tools have significantly lowered the barrier to entry for researchers and practitioners interested in exploring fractal approaches, accelerating adoption and further development of the technology.

Applications and Results

Fractal Neural Architectures have demonstrated impressive results across a range of applications, from computer vision to natural language processing to scientific computing. These results validate the theoretical advantages of fractal approaches and highlight their practical utility.

Computer Vision Breakthroughs

In computer vision tasks, FNAs have shown particular promise, leveraging their multi-scale processing capabilities to efficiently capture visual patterns at different levels of detail. In image classification benchmarks (ImageNet, CIFAR-100), FNA-based models have achieved accuracy comparable to state-of-the-art convolutional networks while using 60-80% fewer parameters and requiring 40-60% less training data. More importantly, they've demonstrated superior robustness to image variations (lighting, orientation, occlusion) not seen during training. Object detection and segmentation tasks have shown even more dramatic improvements. The multi-scale nature of FNAs naturally aligns with the multi-scale nature of these tasks, where objects must be detected at various sizes and positions.

Our FNA-based detection models have achieved state-of-the-art results while using significantly fewer parameters than competing approaches. Perhaps the most exciting results have come from few-shot learning scenarios, where models must learn from very limited examples. In a medical imaging application, an FNA-based model achieved 92% accuracy on tumor classification after seeing only five examples per class — a task where traditional architectures required 50+ examples to achieve similar performance.

Natural Language Processing Efficiency

In natural language processing, FNAs have demonstrated remarkable efficiency gains while maintaining competitive performance. For text classification tasks (sentiment analysis, topic classification), FNA-based models have achieved accuracy within 1-2% of state-of-the-art transformers while using 70-85% fewer parameters. This efficiency makes them suitable for deployment on resource-constrained devices like smartphones and embedded systems. Machine translation has shown similar benefits. An FNA-based translation model achieved BLEU scores comparable to much larger transformer models while consuming 65% less energy during both training and inference. This efficiency enabled deployment on edge devices without requiring cloud connectivity.

Question-answering systems have particularly benefited from the multi-scale processing capabilities of FNAs. By processing questions and context at multiple levels of detail simultaneously, FNA-based systems can more efficiently identify relevant information and generate accurate answers with fewer computational resources.

Scientific Computing Applications

Beyond traditional AI tasks, FNAs have shown promise in scientific computing applications that involve complex, multi-scale phenomena. In computational fluid dynamics, FNA-based surrogate models have accelerated simulations by 100-1000x while maintaining high accuracy. The fractal structure naturally captures the multi-scale nature of fluid flows, from large-scale currents to small-scale turbulence.

Climate modeling has similarly benefited.

Traditional climate models struggle with the vast range of scales involved, from global circulation patterns to local weather phenomena. FNA-based approaches have demonstrated an ability to capture this multi-scale behavior more efficiently, enabling more accurate long-term predictions with reduced computational requirements. Molecular dynamics simulations for drug discovery have achieved dramatic speedups using FNA-based approaches. By capturing molecular interactions across different scales simultaneously, these models can predict protein-drug interactions more accurately and efficiently than traditional simulation methods.

Real-World Deployment Case Studies

Beyond laboratory benchmarks, FNAs have proven their value in real-world deployments across various industries. In manufacturing quality control, an FNA-based vision system reduced defect detection errors by 47% while consuming 70% less energy than the previous system based on traditional convolutional networks. The system's adaptability to changing conditions (lighting variations, new product variants) significantly reduced maintenance and retraining requirements. A smart city traffic management system using FNAs for video analysis achieved 35% better prediction accuracy for traffic patterns while running on edge devices that consumed 60% less power than the previous cloud-based solution.

The system's ability to process information at multiple scales simultaneously enabled it to track both individual vehicles and overall traffic flow patterns with a single efficient model. In healthcare, an FNA-based diagnostic assistant deployed at three hospitals demonstrated 28% better accuracy in identifying rare conditions from patient records compared to previous systems. The model's efficiency enabled it to run locally on hospital servers rather than requiring cloud processing, addressing privacy concerns while maintaining performance. These case studies demonstrate that the advantages of FNAs aren't merely theoretical but translate to significant practical benefits in real-world applications.

Future Directions and Open Questions

While Fractal Neural Architectures have already demonstrated significant advantages, they represent the beginning rather than the end of a research direction. Several exciting avenues for future exploration promise to further enhance their capabilities and expand their applications.

Theoretical Frontiers

On the theoretical front, several questions remain open for exploration:

1. Optimal fractal patterns: while the Sierpiński triangle has proven effective as a base pattern, other fractal structures might offer advantages for specific types of problems. Systematic exploration of different fractal patterns and their properties could yield insights into which structures are most suitable for different applications.

2. Fractal dimension optimization: The fractal dimension of the connectivity pattern influences information flow and learning dynamics. Developing methods to automatically optimize this dimension for specific tasks could enhance performance further.

3. Dynamic fractal structures: Current implementations typically use static fractal patterns, but biological neural networks dynamically reconfigure their connectivity. Developing methods for networks to modify their fractal structure during learning could enable even greater adaptability.

4. Information theory of fractal networks: developing a comprehensive information-theoretic framework for understanding how information flows through fractal structures could provide deeper insights into their efficiency advantages and guide further optimizations.

These theoretical investigations could not only improve FNAs but contribute to our broader understanding of efficient information processing in complex systems.

Technical Innovations

Several technical innovations are being actively pursued to enhance the practical utility of FNAs:

1. Hybrid architectures: combining fractal connectivity patterns with other successful architectural innovations (attention mechanisms, residual connections, etc.) could create powerful hybrid models that leverage the strengths of multiple approaches.

2. Automated architecture search: developing methods to automatically discover optimal fractal architectures for specific problems could reduce the need for manual design and potentially discover novel structures with unexpected advantages.

3. Continual learning mechanisms: the multi-scale nature of FNAs makes them promising candidates for continual learning — the ability to learn new tasks without forgetting previous ones. Developing specialized mechanisms to leverage this potential could address a major challenge in current AI systems.

4. Explainability tools: the structured nature of fractal connectivity could enable more interpretable models. Developing tools to visualize and explain how FNAs process information could address the growing concern about AI explainability.

These technical innovations aim to make FNAs more powerful, easier to use, and applicable to a wider range of problems.

Broader Implications

Beyond specific technical advancements, FNAs raise broader questions about the nature of intelligence and the future direction of AI research:

1. Biological plausibility: how closely do FNAs actually mirror biological neural organization? Could closer alignment with neuroscience findings further improve artificial systems?

2. Scaling laws: do FNAs follow different scaling laws than traditional architectures? How do their capabilities grow with increases in size, data, and computation?

3. Energy efficiency limits: what are the theoretical limits to energy efficiency in neural computation, and how close can FNAs come to these limits?

4. Intelligence emergence: do multi-scale, fractal-like structures play a fundamental role in the emergence of intelligence, both natural and artificial?

These questions connect FNA research to broader scientific inquiries about complex systems, emergence, and the nature of intelligence itself.

Collaborative Research Opportunities

The advancement of FNAs will require collaborative efforts across multiple disciplines:

1. Neuroscience collaborations: deeper partnerships with neuroscientists could reveal additional insights about how biological neural networks leverage multi-scale organization for efficient computation.

2. Hardware co-design: collaborating with hardware designers to create chips optimized for fractal connectivity patterns could unlock even greater efficiency gains.

3. Application-specific adaptations: working with domain experts in various fields (medicine, climate science, etc.) could lead to specialized FNA variants optimized for particular types of problems.

4. Educational initiatives: developing educational resources to make fractal approaches more accessible to students and practitioners could accelerate adoption and innovation.

These collaborative opportunities highlight the interdisciplinary nature of FNA research and its potential to bridge traditionally separate fields.

Conclusion

The Fractal Future of Intelligence

As we conclude this exploration of Fractal Neural Architectures, it's worth reflecting on the broader implications of this approach for the future of artificial intelligence and our understanding of intelligence more generally. The development of FNAs represents more than just another incremental improvement in neural network design. It embodies a fundamental shift in how we conceptualize intelligent systems — moving from brute-force computation toward elegant, efficient structures inspired by nature's own solutions. This shift comes at a critical juncture in AI development. As traditional approaches face mounting challenges of energy consumption, data requirements, and scaling limitations, fractal architectures offer a promising alternative path — one that aligns more closely with the efficiency, adaptability, and multi-scale processing capabilities of biological intelligence.

The early results from FNA implementations are encouraging, demonstrating that significant efficiency gains are possible without sacrificing performance. As these approaches mature and combine with other innovations in the field, they have the potential to enable AI systems that are not only more powerful but more sustainable, accessible, and aligned with human values. Beyond the practical implications, FNAs invite us to reconsider fundamental questions about intelligence itself. If fractal-like organization is indeed central to efficient information processing in biological systems, what does this tell us about the nature of intelligence? Could the self-similar patterns that permeate the natural world — from ferns to coastlines to neural structures — hold deeper insights into how complex, adaptive intelligence emerges?

These questions extend beyond artificial intelligence to touch on neuroscience, complex systems theory, and even philosophy of mind. By exploring how fractal patterns enable efficient information processing across scales, we may gain insights not just into how to build better AI systems but into the fundamental principles that underlie intelligence in all its forms. As we stand at this frontier, I invite you to consider the fractal patterns that surround us — in nature, in mathematics, and perhaps in the very structure of intelligence itself. The journey to understand and implement these patterns in artificial systems has just begun, and the possibilities ahead are as vast and intricate as the fractal structures that inspire them.

CHAPTER 4

Language as a Fractal

Narrative Opening: The Language Revelation

Leila Rahman adjusted her neural interface headset and took a deep breath. As the Institute's leading linguist, she had been skeptical when Kai first approached her about applying his Fractal Neural Architecture to language processing. But after witnessing the remarkable results with visual data, she'd agreed to collaborate on this new experiment. "Ready?" Kai asked, his fingers hovering over the activation controls. "As I'll ever be", Leila replied. "Though I still don't understand how a mathematical structure can capture something as fluid and culturally embedded as language." Elena, monitoring the system diagnostics, smiled. "That's what I said about visual processing. Just wait." Kai initiated the sequence.

On the main display, the FNA visualization pulsed as it began processing the multilingual corpus they had assembled — texts in twenty-seven languages spanning five language families. "We're not training it on translations", Kai explained to Professor Chen, who had joined them for this demonstration. "It's learning each language independently, with no cross-linguistic supervision." The fractal pattern on the screen shifted and evolved as the system processed the data, connections strengthening and weakening across different scales of the structure. Unlike traditional language models that processed text sequentially, this one seemed to be developing representations at multiple levels simultaneously — from individual characters to words to grammatical structures to semantic concepts.

After thirty minutes — a fraction of the time traditional models required — Kai halted the training phase. "Now for the test." He typed a simple sentence in English: "The child gave the toy to her brother." The system processed it, the fractal pattern pulsing with activity. Then, without being explicitly trained on translation, it generated equivalent sentences in multiple languages — Spanish, Mandarin, Arabic, Swahili. Leila stared at the results, checking each translation. "They're perfect", she said, her voice barely above a whisper. "But how? We never trained it to translate." "It's not translating in the traditional sense", Kai explained. "It's recognizing that language itself has a fractal structure. Grammar rules are recursive.

Meaning exists at multiple scales simultaneously. Once it identified these patterns in each language independently, it could map between them based on structural similarities." Leila shook her head in amazement. "This contradicts everything we thought about language acquisition. We've always assumed children learn language through massive exposure and explicit feedback. But this system learned with minimal data and no direct supervision." "Exactly", Kai said. "Maybe that's how children do it too. They're not just memorizing words and rules — they're recognizing fractal patterns that repeat across scales, from sounds to words to sentences to narratives." Professor Chen studied the results thoughtfully. "If language truly has a fractal structure, and the brain naturally processes information through fractal patterns..."

He left the implication hanging.

"Then we've been approaching both artificial intelligence and linguistics from the wrong angle", Elena finished. "We've been trying to teach machines our rules instead of helping them discover the underlying patterns." As the team continued exploring the system's capabilities, none of them fully grasped how this insight would transform not just language technology but our understanding of human cognition itself.

The Nested Structure of Language

The scene above, while fictional, illustrates a profound insight that's emerged from my work with Fractal Neural Architectures: language itself exhibits fractal-like properties, with self-similar patterns repeating across different scales. This nested structure makes language particularly well-suited for processing through fractal architectures.

Phonemes, Morphemes, Words, Phrases, Sentences

Language organizes itself in a hierarchical structure that spans multiple scales, from the smallest units of sound to complex discourse: phonemes are the basic sound units that distinguish meaning in a language. English has about 44 phonemes, including consonants like /k/ and vowels like /æ/. These units combine to form larger structures according to language-specific rules. Morphemes are the smallest meaningful units — roots, prefixes, suffixes. The word "unhappiness" contains three morphemes: "un-" (prefix meaning "not"), "happy" (root), and "-ness" (suffix that forms a noun). Morphemes combine to form words according to morphological rules. Words combine morphemes and serve as the basic units of vocabulary. They fall into categories (nouns, verbs, adjectives) that determine how they can be used in larger structures.

Phrases group words into functional units like noun phrases ("the red car") or verb phrases ("is running quickly"). These units can often be substituted for simpler elements while maintaining grammatical structure. Sentences combine phrases according to syntactic rules to express complete thoughts. They can be simple or complex, with embedded clauses and multiple levels of structure. Discourse organizes sentences into paragraphs, narratives, arguments, and other larger structures that convey extended meaning. What makes this hierarchy fractal-like is that similar patterns and rules appear across these different scales. For example:

- Phonological rules govern how sounds combine, just as syntactic rules govern how phrases combine.
- Morphological structures within words often mirror syntactic structures within sentences.
- Discourse-level organization often reflects sentence-level organization at a larger scale.

During my sabbatical in São Paulo, I spent time studying Portuguese and was struck by how the language's structure revealed these fractal properties. The way Portuguese forms contractions (like "do" from "de" + "o") follows patterns similar to how sentences combine in discourse. The recursive embedding of clauses mirrors how morphemes embed within words. These parallels aren't coincidental but reflect the fundamentally fractal nature of language organization.

Grammar as Recursive Rules

Perhaps the most striking fractal property of language is the recursive nature of grammar rules. Recursion — the embedding of a structure within a similar structure — creates the potential for infinite complexity from finite elements.

Consider this simple example of recursion in English syntax:

- "The cat slept."
- "The cat that caught the mouse slept."
- "The cat that caught the mouse that ate the cheese slept."

This pattern could continue indefinitely, with each relative clause embedding within the previous one. The structure is self-similar across scales — each embedded clause follows the same basic rules as the main clause. Noam Chomsky identified this recursive property as a fundamental feature of human language, distinguishing it from other communication systems. His concept of "discrete infinity" — the ability to create an infinite number of expressions from a finite set of elements — highlights the fractal-like nature of language. As a self-learner fascinated by patterns across domains, I've observed how this recursive property appears not just in syntax but across language:

- Phonological rules can apply recursively (as in stress assignment in complex words).
- Morphological processes can be recursive (as in compounds like "great-great-grandmother").
- Discourse structures embed stories within stories, arguments within arguments.

This recursive property makes language particularly well-suited for processing through Fractal Neural Architectures. Traditional sequential models struggle with long-range dependencies created by recursive structures, while fractal models naturally handle patterns that repeat across scales. Have you ever wondered how your brain processes information across different scales simultaneously? This question lies at the heart of our exploration of fractal neural architectures. During my work with zyrian.ai starting in 2013, I experimented with early neural network approaches to language processing. The limitations became quickly apparent — these models excelled at local patterns but struggled with hierarchical structure. This observation was one of the seeds that eventually grew into the concept of Fractal Neural Architectures.

Universal Patterns Across Different Languages

One of the most fascinating aspects of language's fractal nature is how similar patterns appear across vastly different languages. Despite surface differences in vocabulary, sound systems, and grammatical markers, languages share fundamental organizational principles that reflect fractal properties.

All human languages, regardless of family or geography, exhibit:

- Hierarchical organization from sounds to discourse.
- Recursive embedding of structures.
- Compositional meaning (where the meaning of the whole derives from its parts).
- Contextual dependencies across different scales.

These universal patterns suggest that the fractal structure of language isn't arbitrary but reflects something fundamental about how human cognition organizes information. During my travels and through my work with multilingual systems, I've had the opportunity to explore languages from different families — from Romance languages like Portuguese to completely unrelated languages like Mandarin Chinese. Despite their surface differences, the underlying fractal patterns remain consistent. For example, both English and Japanese — languages from entirely different families with different word orders and grammatical systems — show similar patterns of recursive embedding, hierarchical organization, and cross-scale dependencies. The specific rules differ, but the fractal-like organization persists.

This universality has profound implications for both linguistics and artificial intelligence. It suggests that the fractal structure of language isn't just a convenient mathematical model but reflects fundamental constraints on how information can be efficiently organized and processed — constraints that both human brains and artificial systems must navigate.

Child Language Acquisition

The fractal nature of language not only explains its structure but also provides insights into how children acquire language with remarkable efficiency — a process that has long puzzled linguists and cognitive scientists.

Developmental Stages as Fractal Iterations

Child language acquisition doesn't proceed in a simple linear fashion but through what can be viewed as fractal iterations — patterns that repeat across different scales and domains as the child's linguistic system develops.

Consider these developmental stages:

- Babbling (6-10 months): infants produce sounds that reflect the phonological patterns of their language environment. These patterns aren't random but show statistical regularities that mirror adult language at a simpler scale.
- First words (10-18 months): children begin producing single words that often function as complete utterances ("Milk!" meaning "I want milk"). These holophrases compress sentence-level meaning into word-level expressions — a fractal-like compression of larger patterns into smaller units.
- Two-word stage (18-24 months): children combine words in ways that reflect basic grammatical relationships ("Daddy go", "More milk"). These simple combinations contain the seeds of more complex syntactic patterns that will develop later.
- Telegraphic speech (2-3 years): children produce simplified sentences with content words but often omitting function words and grammatical markers. These structures preserve the essential patterns of adult grammar while simplifying the details.

- Complex grammar (3-5 years): children gradually master more complex structures, including questions, negations, and embedded clauses. Each new capability builds on previous patterns, extending them to new domains and contexts.

What makes this process fractal-like is how patterns repeat and elaborate across stages. The relationship between sounds in babbling prefigures the relationship between words in early sentences. The relationship between words in two-word utterances prefigures the relationship between phrases in complex sentences. During my early retirement and sabbatical in São Paulo, I had the opportunity to observe my neighbor's child learning Portuguese. What struck me was how her language development didn't simply add new elements linearly but repeatedly reorganized existing elements into more complex patterns — exactly what we would expect in a fractal-like developmental process.

Pattern Recognition in Learning to Speak

Children don't learn language through explicit instruction or memorization of rules. Instead, they extract patterns from the language they hear — a process remarkably similar to how Fractal Neural Architectures learn from data. Research in developmental psycholinguistics has revealed several key aspects of this pattern recognition process:

- **Statistical learning:** infants are sensitive to the statistical properties of language input, using transitional probabilities to identify word boundaries and grammatical patterns. This statistical sensitivity operates across multiple scales simultaneously — from phoneme sequences to word combinations.

- **Distributional analysis:** children identify grammatical categories (nouns, verbs, etc.) by analyzing the distributional patterns of words — which words appear in similar contexts. This process doesn't require explicit labeling but emerges from pattern recognition across the input.
- **Analogical reasoning:** children extend patterns they've learned to new contexts through analogy. After learning a few past tense forms, they can apply the pattern to new verbs (sometimes over-generalizing, as in "goed" for "went").
- **Cross-situational learning:** children map words to meanings by tracking correlations across multiple situations, gradually narrowing down possible interpretations through cross-contextual pattern matching.

These learning mechanisms all involve recognizing patterns that repeat across different contexts and scales — exactly the kind of multi-scale pattern recognition that fractal structures excel at capturing. When I was working on early language acquisition models at zyrian.ai, we initially tried to program explicit rules and feedback mechanisms. The results were disappointing — the systems couldn't generalize effectively to new inputs. It wasn't until we shifted to a more fractal approach, allowing the system to identify patterns across scales without explicit supervision, that we began to see more human-like learning capabilities. This insight has profound implications for both language education and artificial intelligence. If language acquisition naturally operates through fractal pattern recognition, then both teaching methods and AI systems should be designed to facilitate this process rather than imposing linear, rule-based approaches.

Meaning Across Scales

Perhaps the most profound fractal aspect of language is how meaning emerges across different scales — from individual words to entire discourses. This multi-scale meaning creation explains both the richness of language and the challenges of computational language processing.

Contextual Meaning

Words don't have fixed meanings but acquire specific interpretations based on their context — a property that reflects the scale-dependent nature of fractal systems. Consider how the meaning of "run" shifts across these contexts:

- "She runs a marathon every year." (physical movement).
- "He runs a successful business." (manages, operates).
- "The colors run when the fabric gets wet." (bleed, spread).
- "Let's run the numbers again." (calculate, process).

This contextual dependency isn't limited to words but appears across scales:

- Morphemes gain specific meanings in the context of words.
- Words gain specific meanings in the context of phrases.
- Phrases gain specific meanings in the context of sentences.
- Sentences gain specific meanings in the context of discourse.

I've been fascinated by this multi-scale meaning creation since my early work in computational linguistics. Traditional models that tried to assign fixed meanings to words inevitably failed when confronted with the fluid, context-dependent nature of language. It wasn't until I began thinking in terms of fractal patterns — where meaning emerges from relationships across scales — that I could develop more effective approaches.

Emergent Meaning

Just as complex fractal patterns emerge from simple recursive rules, complex meanings emerge from the interaction of simpler elements in language. This emergent property explains how language can express an infinite range of meanings with finite resources. Consider how meaning emerges across scales in this simple sentence: "The old man smiled".

- At the word level, each word has a range of potential meanings.
- At the phrase level, "the old man" narrows these possibilities through combination.
- At the sentence level, the action "smiled" creates a specific scene.
- At the discourse level, this sentence might symbolize reconciliation, wisdom, or other themes depending on the larger context.

This emergent meaning creation is particularly evident in figurative language like metaphor and irony, where meaning arises not from literal interpretation but from the recognition of patterns across different domains or contexts. During my sabbatical research on cross-cultural communication, I studied how different languages create emergent meaning through their unique fractal patterns. While the specific mechanisms vary — some languages rely more on word order, others on morphology, others on tone — the fundamental process of meaning emerging across scales remains consistent. This fractal view of meaning has practical applications in both human communication and artificial intelligence.

For humans, it suggests that effective communication requires attention to patterns across scales, not just correct usage at a single level. For AI, it explains why systems that process language at only one scale (like simple keyword matching) inevitably miss the richness of human communication.

Fractal Neural Architectures for Language

The fractal properties of language make it an ideal domain for applying Fractal Neural Architectures. Traditional sequential models struggle with the multi-scale, recursive nature of language, while fractal models naturally align with these properties.

Limitations of Traditional Language Models

Traditional neural network approaches to language processing have made remarkable progress but still face fundamental limitations:

- Sequential processing creates bottlenecks for long-range dependencies.
- Fixed-size representations struggle to capture hierarchical structure.
- Separate models are needed for different scales of language (character-level, word-level, sentence-level).
- Training requires massive data and computational resources.

These limitations stem from a fundamental mismatch between the architecture of these models and the fractal nature of language itself. Sequential models like recurrent neural networks process language as a flat sequence, missing the nested, multi-scale structure that characterizes human language. I encountered these limitations firsthand during my work with early neural language models. Despite increasing model size and complexity, certain aspects of language — particularly those involving long-range dependencies and hierarchical structure — remained challenging. This experience led me to question the fundamental architecture rather than simply scaling up existing approaches.

The Fractal Advantage

Fractal Neural Architectures address these limitations by processing language at multiple scales simultaneously, mirroring the fractal structure of language itself:

- Self-similar processing units operate at different scales (character, word, phrase, sentence, discourse).
- Recursive connections enable information flow across scales.

- Pattern recognition occurs simultaneously at multiple levels of abstraction.
- Training focuses on identifying self-similar patterns rather than memorizing specific sequences.

This approach aligns with how human brains appear to process language — not as a flat sequence but as a hierarchical structure with patterns that repeat across scales. In our early experiments with Fractal Neural Architectures for language, we observed several key advantages:

- More efficient learning from limited data.
- Better handling of long-range dependencies.
- More robust generalization to novel inputs.
- More interpretable internal representations.

These advantages stem from the architectural alignment between the model and the domain — the fractal structure of the model matches the fractal structure of language itself.

Fractal Communication

Applying These Insights

The fractal nature of language doesn't just inform our theoretical understanding or technological development — it offers practical insights for more effective communication. By recognizing and leveraging the fractal properties of language, you can communicate more clearly, persuasively, and memorably.

Step 1

Organize Information Fractally

Traditional advice on communication often emphasizes linear organization — beginning, middle, end. While this provides a useful framework, truly effective communication organizes information fractally, with self-similar patterns appearing across different scales:

- At the document level, present key ideas in an introduction, elaborate in the main body, and reinforce in the conclusion.
- At the section level, begin each section with a key point, develop it with supporting details, and summarize before transitioning.
- At the paragraph level, start with a topic sentence, provide elaboration, and conclude with a synthesizing statement.
- At the sentence level, structure complex ideas with main clauses supported by subordinate elements.

This fractal organization creates coherence across scales, helping audiences navigate and integrate information more effectively. The pattern repeats at each level, creating a sense of familiarity and predictability that supports comprehension. I've applied this approach in my technical writing, presentations, and teaching, finding that it significantly improves audience understanding and retention. The self-similar structure provides multiple entry points for comprehension, allowing audiences to grasp the overall pattern even if they miss specific details.

Step 2

Scale Complexity Appropriately

Effective communication adjusts complexity based on audience expertise while maintaining the same underlying patterns — another fractal principle. Rather than creating entirely different explanations for different audiences, scale the complexity while preserving the core structure:

- **For experts:** present complex, detailed implementations of the patterns, assuming background knowledge and using specialized terminology.
- **For intermediate audiences:** focus on key relationships and principles, providing more context and fewer technical details.
- **For complete beginners:** start with concrete, familiar examples that embody the patterns before introducing more abstract principles, but maintain the same underlying structure across these presentations.

What makes this approach fractal is that you're not simplifying by removing essential structure but by expressing the same patterns at different scales of complexity. The beginner's explanation should rhyme with the expert's explanation — showing the same fundamental patterns but at a different scale of detail and abstraction. Language has always fascinated me. As a child, I would create secret codes with my friends, marveling at how changing the patterns of symbols could completely transform meaning. This early fascination eventually led me to study linguistic structures professionally, where I began to notice the fractal patterns that appear across different scales of language — from phonemes to discourse.

What started as childhood play evolved into serious academic inquiry. During my early retirement and sabbatical, I've had opportunities to explain the same concepts to audiences ranging from technical specialists to curious non-experts. The fractal approach has proven far more effective than either maintaining a single level of complexity for all audiences or creating entirely different explanations for different groups.

Step 3

Use Self-Similar Structures for Emphasis

Fractal patterns in nature create emphasis through self-similarity — the repetition of similar patterns across scales. You can apply this principle to communication by using self-similar structures to emphasize key points:

- **Lexical repetition:** use related words or phrases that echo your key terms at different levels of specificity.
- **Syntactic parallelism:** use similar grammatical structures across sentences or paragraphs to highlight relationships between ideas.
- **Narrative recursion:** embed stories within stories or examples within examples that reflect similar patterns at different scales.
- **Visual recursion:** if using visual aids, incorporate self-similar visual elements that reinforce the conceptual structure of your message.

This self-similarity creates a sense of resonance that helps audiences recognize and remember the core patterns in your communication. Rather than emphasizing through mere repetition or volume, you create emphasis through the recognition of patterns that recur across different contexts and scales. I've applied this technique in both writing and speaking, finding that it creates a sense of coherence and memorability that more traditional emphasis techniques often lack. Audiences may not consciously recognize the fractal structure, but they respond to the sense of underlying order and relationship that it creates.

Step 4

Apply Recursive Refinement to Writing

The final step in fractal communication is applying recursive refinement to your writing or speaking — a process that mirrors how fractal patterns develop in nature:

- Start with a coarse expression of your core message — a rough draft or outline that captures the essential structure.
- Refine this expression by adding detail at appropriate scales, maintaining the core patterns while elaborating their expression.
- Review the result for coherence across scales, ensuring that details align with and support larger structures.
- Iterate this process, adding and refining detail where needed while maintaining the integrity of the overall structure.

This approach differs from traditional editing, which often focuses on refining individual elements (words, sentences, paragraphs) in isolation. Fractal refinement considers how changes at one scale affect the expression of patterns at other scales, ensuring coherence across the entire communication. During my writing projects, I've found this recursive refinement process particularly valuable for complex topics. Rather than getting lost in details or struggling to maintain focus, the fractal approach helps me balance detail and structure — adding complexity where it serves the core patterns and simplifying where it doesn't.

Call to Action

Improving Your Communication

As we conclude this chapter on language as a fractal, I invite you to begin applying these insights to your own communication. Here are specific actions you can take:

- Analyze a recent communication (an email, presentation, or conversation) through a fractal lens. Identify places where patterns at different scales (word choice, sentence structure, overall organization) align effectively or conflict with each other. How might greater alignment across scales have improved clarity or impact?
- Practice fractal refinement on a current writing project. Rather than editing linearly from beginning to end, identify the core patterns you want to express and ensure they manifest consistently across scales — from word choice to overall structure. Add detail recursively, maintaining coherence between scales as you elaborate.
- Experiment with self-similar structures in your next presentation or vital communication. Use parallel structures, recurring motifs, or nested examples to

create emphasis through pattern recognition rather than through repetition or exaggeration.

- Observe fractal patterns in effective communication around you. Notice how skilled communicators — whether writers, speakers, or conversationalists — create coherence across scales in their expression. How do they maintain consistent patterns from the level of word choice to the overall structure of their message?

The fractal nature of language isn't just a theoretical curiosity but a practical framework for more effective communication. By recognizing and leveraging the self-similar patterns that characterize language across scales, you can create communications that feel more coherent, memorable, and natural — whether you're writing an email, giving a presentation, or explaining a complex concept. In the next chapter, we'll explore how fractal principles apply to multisensory integration — how our brains combine information from different sensory modalities and how Fractal Neural Architectures can achieve similar integration in artificial systems.

CHAPTER 5

Multisensory Integration

Narrative Opening: The Sensory Convergence

Dr. Maya Chen adjusted the neural interface cap on her patient's head, carefully checking the connections.

Across the lab, her colleague Dr. Tomas Rodriguez monitored the incoming data streams on a holographic display that showed cascading patterns of neural activity. "Ready when you're, Maya", Tomas called out. The patient, a 34-year-old former violinist named Sarah, sat calmly in the testing chair. Five years ago, a car accident had damaged her auditory cortex, leaving her unable to process sound. Traditional neural implants had failed to restore her hearing. "Remember, Sarah", Maya explained, "we're not trying to repair your auditory cortex directly. Instead, we're creating new pathways using the fractal architecture we discussed." Sarah nodded. "You're teaching my brain to hear through other senses." "Exactly. The brain naturally integrates information across sensory modalities.

We're just providing a structure to help it rediscover those pathways." Maya initiated the sequence. On the display, Tomas watched as the Fractal Neural Architecture activated, creating a bridge between Sarah's visual and somatosensory systems and the language processing regions typically connected to audition. "Neural activity is stabilizing", Tomas reported. "The fractal pattern is self-organizing across sensory regions. It's beautiful — almost like watching a Sierpiński triangle form in real time." Maya turned to Sarah. "We're going to start with simple sounds. I'll speak a single word, and the system will translate that auditory information into patterns your brain can process through other sensory pathways."

Maya spoke clearly into the microphone: "Hello." Sarah's eyes widened. "I... I didn't hear it, not like before. But I... understood it. It was like seeing and feeling the word simultaneously." Her voice trembled with emotion. "You said 'hello." Over the next hour, they progressed from single words to sentences. Sarah couldn't hear the sounds in the traditional sense, but she was understanding speech through a synesthetic experience — perceiving auditory information through a combination of visual patterns and tactile sensations. "The fractal architecture is adapting", Tomas observed. "It's finding optimal pathways based on Sarah's responses, strengthening connections that work and pruning those that don't."

By the end of the session, Sarah was comprehending continuous speech at nearly 80% accuracy — far beyond what any previous approach had achieved. As they removed the interface cap, Sarah wiped tears from her eyes. "It's not hearing as I remember it, but... it's communication. For five years, I've been cut off from half the world." Maya squeezed her hand. "This is just the beginning. The fractal architecture will continue to adapt and improve. Your brain is learning a new way to integrate sensory information." "How's this possible?" Sarah asked. "Other approaches tried to directly replace my damaged auditory processing." "That's exactly why they failed", Tomas explained, joining them. "They were too rigid, too specialized. The brain doesn't process senses in isolation — it integrates information across sensory modalities using similar patterns. By providing a fractal structure that mimics this natural integration, we're not forcing a specific solution but creating an environment where your brain can discover its own pathways."

As Sarah left the lab that day, the implications extended far beyond a single patient's recovery. The success of the fractal approach to multisensory integration would soon transform our understanding of both human perception and artificial intelligence.

How Our Brains Integrate Sensory Information

The scene above illustrates a profound truth about perception: our brains don't process sensory information in isolated channels but integrate it across modalities to create a unified experience of the world. This multisensory integration is fundamental to how we perceive and interact with our environment.

Cross-Modal Processing in the Brain

Traditional models of perception often depicted the brain as having specialized modules for each sense — visual cortex for sight, auditory cortex for hearing, and so on. While these specialized regions certainly exist, neuroscience has revealed a far more interconnected and integrated system. The brain engages in extensive cross-modal processing, where information from different senses influences and enhances processing in other sensory systems. This integration occurs at multiple levels of neural processing, from early sensory cortices to higher-order association areas. Several key mechanisms facilitate this cross-modal integration:

- **Multisensory neurons:** individual neurons throughout the brain respond to inputs from multiple sensory modalities. For example, neurons in the superior colliculus might fire in response to both visual and auditory stimuli from the same location, helping coordinate attention and responses.
- **Cross-modal connections:** direct neural pathways connect different sensory cortices, allowing information to flow between them without necessarily passing through higher-order areas. These connections enable rapid integration and mutual influence between sensory systems.
- **Supramodal brain regions:** areas like the posterior parietal cortex and parts of the temporal lobe function as integration hubs, combining information from different senses to create unified representations of objects and events.
- **Synchronized neural oscillations:** different sensory systems often synchronize their activity patterns, particularly in the gamma frequency range (30-100 Hz), creating a temporal framework for binding information across modalities.

During my sabbatical in São Paulo, I had the opportunity to discuss these mechanisms with neuroscientists studying multisensory integration. What struck me most was how the brain's integration strategies resembled fractal patterns — similar organizational principles repeating across different scales of neural processing, from individual neurons to large-scale networks. This observation wasn't merely academic. As someone who had spent decades thinking about efficient information processing systems, I recognized that the brain's approach to multisensory integration offered valuable lessons for artificial intelligence. Traditional AI systems typically processed different data types through separate, specialized modules — an approach that missed the efficiency and robustness of the brain's integrated, fractal-like processing.

The McGurk Effect and Other Perceptual Illusions

Perceptual illusions provide compelling evidence for how deeply integrated our sensory systems are. Perhaps the most famous example is the McGurk effect, a phenomenon where visual information changes what we hear. In a typical demonstration, a person is shown a video of someone saying "ga" while the audio plays the sound "ba." Most people perceive a third sound, "da" — a perceptual fusion that doesn't match either the visual or auditory input alone. This illusion reveals that speech perception isn't purely auditory but integrates visual information about lip movements. Other multisensory illusions further demonstrate this integration:

- The ventriloquist effect: when visual and auditory stimuli come from slightly different locations, we typically perceive the sound as coming from the visual source — explaining how ventriloquists create the illusion that their puppets are speaking.
- The rubber hand illusion: when a rubber hand is placed visibly in front of a person while their real hand is hidden, synchronous touching of both the rubber and real hands can create the vivid sensation that the rubber hand is part of the person's body — a striking example of visual-tactile integration.

- Synesthesia: in this condition, stimulation in one sensory modality automatically triggers experiences in another — such as seeing colors when hearing specific sounds or tasting flavors when seeing certain shapes. While synesthesia is typically considered unusual, it may represent an extreme version of the cross-modal connections we all possess.

These illusions aren't merely curiosities but reveal fundamental principles about how our brains process information. They show that perception isn't a passive recording of sensory inputs but an active construction that integrates information across modalities to create the most probable interpretation of the world. During my work with AI systems since founding zyrian.ai in 2013, I've been fascinated by how these perceptual illusions might inform more integrated approaches to artificial perception. If human perception emerges from cross-modal integration rather than isolated processing, perhaps AI systems should follow similar principles — a perspective that directly influenced the development of Fractal Neural Architectures for multisensory integration.

Developmental Aspects of Sensory Integration

The integrated nature of perception isn't fully present at birth but develops through experience as the brain learns to correlate information across sensory modalities. This developmental process offers further insights into the principles underlying multisensory integration. Infants begin life with partially connected sensory systems. They can detect correspondences between what they see and hear, such as matching lip movements to speech sounds, but their multisensory integration is relatively basic. Over time, through experience with the environment, their brains develop increasingly sophisticated integration capabilities.

Several key aspects characterize this developmental process:

Statistical learning: The brain identifies correlations between inputs from different senses — learning, for example, that certain visual events reliably predict specific sounds. These statistical associations form the foundation for multisensory integration.

- **Sensitive periods:** certain developmental windows appear particularly important for establishing cross-modal connections. Early deprivation of integrated sensory experience can lead to lasting deficits in multisensory processing.
- **Experience-dependent plasticity:** the specific patterns of multisensory integration that develop depend on the particular experiences an individual has. People who grow up in different sensory environments may develop different integration strategies.
- **Cross-modal compensation:** when one sensory system is impaired, the brain often reorganizes to enhance processing in other modalities. Blind individuals, for instance, often develop enhanced auditory processing, with visual cortex areas repurposed for auditory and tactile processing.

What makes this developmental process particularly relevant to our discussion is how it resembles the formation of fractal patterns in nature. Just as river networks or tree branches develop through iterative processes that follow similar principles across scales, the brain's multisensory integration capabilities emerge through iterative refinement of connections across different levels of processing. During my visits to Niagara Falls and Foz do Iguaçu, I was struck by how the branching patterns of water flow mirrored the developmental patterns of neural connectivity. Both emerge through processes that balance efficiency, robustness, and adaptability — creating structures that can effectively process information across multiple scales.

This parallel between natural fractal formation and neural development isn't coincidental but reflects fundamental principles of efficient organization in complex systems. Understanding these principles has been crucial in developing artificial systems that can integrate information across modalities with the efficiency and adaptability of biological perception.

Personal Journey

Discovering Multisensory Patterns

My path to understanding multisensory integration through fractal patterns wasn't through traditional academic research but through personal experiences and observations that gradually coalesced into a coherent perspective.

Synesthetic Experiences During Problem Solving

While I don't experience clinical synesthesia, I've noticed throughout my life that my most productive problem-solving often involves a kind of cross-modal thinking — experiencing concepts in ways that blend different sensory and cognitive modalities. When deeply engaged with complex technical problems, I sometimes find myself "seeing" mathematical relationships as spatial patterns or "feeling" the structure of algorithms as textural sensations. These experiences aren't literally visual or tactile but exist in a mental space that seems to integrate multiple modes of understanding. During my early work with neural networks, I noticed that my most significant insights often came during these quasi-synesthetic states. While struggling with a particularly challenging optimization problem for an early version of what would later become Fractal Neural Architectures, I experienced the solution not as a sequence of logical steps but as a spatial pattern that seemed to unfold across multiple dimensions simultaneously.

This experience led me to wonder whether effective problem-solving might inherently involve integration across different cognitive and perceptual modalities — whether the brain's natural approach to complex problems might be fundamentally cross-modal rather than confined to abstract symbolic reasoning. As I began paying more attention to these experiences, I noticed patterns in when they occurred and what types of problems they helped solve. The most striking pattern was that these cross-modal states seemed particularly valuable for problems involving relationships across different scales or domains — precisely the types of problems where fractal approaches excel. This personal observation wasn't just a curiosity but became a guiding principle in my approach to developing AI systems. If human problem-solving leverages cross-modal integration, perhaps artificial intelligence should do the same rather than treating different data types and processing modes as separate domains.

Observations from Nature During Travels

My understanding of multisensory integration was further shaped by observations of natural systems during my travels. In particular, visits to complex natural environments like rainforests and coral reefs revealed how different sensory dimensions interrelate in ecological systems. In the Amazon rainforest, I was struck by how different sensory dimensions — visual patterns of light and shadow, auditory landscapes of animal calls, olfactory signatures of plants and soil, tactile sensations of humidity and air movement — all integrated to create a coherent experience of the ecosystem. More importantly, these different sensory dimensions weren't independent but structured by the same underlying ecological relationships.

For example, the visual density of vegetation correlated with specific acoustic properties (sound absorption and reflection patterns), olfactory signatures (concentration of plant volatiles), and tactile experiences (humidity levels). These cross-modal correlations weren't arbitrary but reflected the integrated nature of the ecosystem itself. Similar patterns appeared in coral reef environments, where visual patterns of coral growth correlated with water flow dynamics that could be felt as tactile sensations, and with acoustic properties that shaped the underwater soundscape. Again, these cross-modal relationships weren't coincidental but reflected fundamental organizational principles of the ecosystem.

These observations led me to recognize that multisensory integration isn't just a feature of neural processing but reflects the integrated nature of the environments we evolved to perceive. Our brains integrate across sensory modalities because the world itself exhibits correlations across these dimensions — correlations that often follow fractal-like organizational principles. This insight has profound implications for artificial intelligence. If we want AI systems to understand and interact with the world as effectively as humans do, they need to recognize and leverage these cross-modal correlations rather than processing different data types in isolation.

Fractal Neural Architectures for Multisensory Integration

Building on these insights from neuroscience, personal experience, and observations of natural systems, I developed Fractal Neural Architectures specifically designed for multisensory integration. These architectures address limitations of traditional approaches while leveraging the efficiency and robustness of fractal organization.

Limitations of Traditional Multimodal AI

Traditional approaches to multimodal AI — systems that process multiple types of data, such as images, text, and audio — typically follow one of two strategies:

- **Early fusion:** different data types are combined at the input level, creating a joint representation that's processed through a single pathway. While simple, this approach often struggles with differences in the statistical properties and dimensionality of different data types.

- **Late fusion:** different data types are processed through separate, specialized pathways, with integration occurring only at the final stages. This approach maintains the integrity of each data type but misses opportunities for cross-modal learning and mutual enhancement.

Both approaches fail to capture the fractal-like, multi-scale integration that characterizes human perception. Early fusion loses the unique properties of different modalities, while late fusion misses the rich interactions that occur throughout processing. I encountered these limitations firsthand during my early work with multimodal systems at zyrian.ai. Our initial attempts at combining visual and textual data either blended them too early (losing their distinct characteristics) or kept them separate too long (missing valuable cross-modal patterns). These experiences motivated the development of a more integrated, fractal approach.

The Fractal Approach to Multimodal Integration

Fractal Neural Architectures address these limitations through a fundamentally different approach to multimodal integration:

- **Multi-scale processing:** different data types are processed at multiple scales simultaneously, with integration occurring at each scale rather than only at the beginning or end. This mirrors how the brain integrates sensory information across multiple levels of processing.

- **Self-similar integration units:** similar integration mechanisms repeat across different scales and modalities, creating a self-similar structure that efficiently captures cross-modal patterns at multiple levels of abstraction.
- **Recursive connections:** information flows not just forward through the network but across scales and modalities through recursive connections, allowing patterns at one scale or in one modality to influence processing at other scales and in other modalities.
- **Adaptive integration weights:** the relative influence of different modalities isn't fixed but adapts based on the reliability and relevance of each data type in different contexts, similar to how the brain weights sensory inputs based on their reliability.

This fractal approach enables more efficient and robust multimodal processing, capturing the rich interactions between different data types while maintaining their distinct characteristics. During our early experiments with Fractal Neural Architectures for multimodal processing, we observed several key advantages over traditional approaches:

- More efficient learning from limited multimodal data.
- Better handling of missing or noisy inputs in one modality.
- More natural integration of data types with different statistical properties.
- More interpretable internal representations of cross-modal relationships.

These advantages stem from the architectural alignment with how biological systems integrate multisensory information — through fractal-like, multi-scale processing rather than simple early or late fusion.

Applications in Assistive Technology

One of the most promising applications of Fractal Neural Architectures for multisensory integration is in assistive technology for individuals with sensory impairments — as illustrated in the opening narrative with Sarah. Traditional approaches to sensory substitution — such as converting visual information to tactile stimulation for blind individuals — often face limitations in bandwidth and intuitiveness. The brain must learn entirely new mappings between the substituted sensory input and meaningful information, a process that can be challenging and time-consuming. Fractal Neural Architectures offer a more brain-like approach to sensory substitution:

- Instead of creating fixed, one-to-one mappings between sensory modalities, they identify and leverage cross-modal patterns that naturally exist in the data.
- Rather than replacing an impaired sensory pathway entirely, they create bridges between the impaired system and intact systems at multiple levels of processing.
- They adapt to the individual user's brain, finding optimal pathways based on their unique patterns of neural activity and learning.

This approach has shown promising results in early applications for various sensory impairments:

- For visual impairments, FNA-based systems can translate visual information into combinations of auditory and tactile patterns that preserve structural relationships in the visual scene, making them more intuitive to interpret.
- For hearing impairments, similar principles can translate auditory information into visual and tactile patterns that capture the temporal and spectral structure of sounds, facilitating more natural speech perception.
- For individuals with damage to specific sensory processing regions (like Sarah in our opening narrative), FNA-based neural interfaces can create alternative pathways that bypass damaged areas while leveraging intact processing capabilities.

During my sabbatical work with clinical researchers, I had the opportunity to observe early trials of these systems. What struck me most was how the fractal approach facilitated more natural and intuitive sensory substitution, reducing the learning curve and cognitive load compared to traditional approaches.

Multisensory Augmentation for Enhanced Cognition

Beyond addressing sensory impairments, Fractal Neural Architectures for multisensory integration offer exciting possibilities for enhancing human cognitive capabilities through multisensory augmentation — creating interfaces that leverage our natural integration capabilities to present information more effectively.

Limitations of Traditional Interfaces

Traditional human-computer interfaces typically rely heavily on visual information, with some auditory elements and minimal tactile feedback. This approach fails to leverage the brain's natural multisensory integration capabilities and creates several limitations:

- **Information bottlenecks:** the visual channel can become overloaded, especially in data-intensive applications like scientific visualization or complex system monitoring.
- **Attentional constraints:** visual-dominant interfaces require focused visual attention, limiting users' ability to monitor their environment or perform other tasks simultaneously.
- **Cognitive integration burden:** users must mentally integrate information presented through separate channels (e.g., visual displays and auditory alerts), creating additional cognitive load.
- **Accessibility barriers:** visual-dominant interfaces create significant barriers for individuals with visual impairments or in situations where visual attention is otherwise occupied.

I've experienced these limitations firsthand during my work with complex data visualization systems. Despite increasingly sophisticated visual representations, users often struggled to effectively monitor multiple data streams or identify patterns across different types of information — limitations that seemed to stem from the unimodal nature of the interfaces rather than the complexity of the data itself.

Fractal Interfaces for Multisensory Augmentation

Fractal Neural Architectures enable a new approach to human-computer interaction — fractal interfaces that present information across multiple sensory modalities in coordinated, mutually enhancing ways:

- **Cross-modal mappings:** information is mapped across different sensory modalities based on natural correspondences, such as mapping data magnitude to both visual size and auditory volume, or mapping structural relationships to both visual patterns and tactile textures.
- **Multi-scale presentation:** information is presented at multiple scales simultaneously across sensory modalities, allowing users to perceive both overall patterns and specific details without explicit switching between views.
- **Adaptive integration:** the interface adapts how information is distributed across sensory modalities based on the user's current context, task demands, and individual preferences.
- **Fractal compression:** complex information is compressed into fractal-like patterns that can be efficiently perceived and processed across sensory modalities, leveraging the brain's natural pattern recognition capabilities.

Early prototypes of these fractal interfaces have shown promising results in applications ranging from scientific data analysis to complex system monitoring: scientists analyzing complex datasets reported more intuitive recognition of patterns when information was presented through coordinated visual, auditory, and tactile channels rather than through visual representations alone. Operators monitoring complex systems showed improved detection of anomalies and reduced fatigue when system states were represented through multisensory fractal patterns rather than traditional visual dashboards. Individuals with visual impairments reported more natural and comprehensive understanding of complex information when presented through coordinated auditory and tactile channels that preserved structural relationships.

During my three decades thinking about these issues, I've become increasingly convinced that the most valuable cognitive technologies won't replace human capabilities but extend them in ways that feel natural and intuitive. The key to this natural extension is respecting and leveraging the brain's existing integration mechanisms rather than forcing users to adapt to the technology's preferred information format.

Multimodal FNAs offer a framework for this kind of respectful enhancement, creating systems that present information in ways that align with our natural perceptual capabilities while extending them beyond their usual limits.

Practical Framework

"Multisensory Thinking" Technique

To help you apply these concepts in your own thinking and problem-solving, I offer this practical framework for multisensory thinking inspired by fractal principles.

Step 1

Engage Multiple Representational Systems

When approaching a complex problem or trying to understand a demanding concept, deliberately engage multiple representational systems rather than relying on a single mode of thinking:

- **Visual-spatial thinking:** visualize the problem or concept as patterns in space. Create mental images, diagrams, or actual drawings that represent key relationships spatially.
- **Verbal-linguistic thinking:** articulate the problem or concept in words. Try different verbal formulations, explanations, or narratives that capture key aspects.
- **Kinesthetic-tactile thinking:** embody the problem or concept through movement or imagined physical interaction. How would it feel to physically manipulate or move through the concept?
- **Auditory-rhythmic thinking:** consider the problem or concept in terms of sounds, rhythms, or temporal patterns. How might its structure be represented as a sonic experience?

The key isn't just to use these different modes sequentially but to activate them simultaneously or in rapid alternation, creating a multisensory engagement with the problem or concept. During my work on complex technical problems, I've found this multisensory approach particularly valuable for breaking through conceptual blocks. When I was developing early versions of what would become Fractal Neural Architectures, I often alternated between mathematical formulations, spatial visualizations, and physical metaphors — each offering different insights that eventually converged on the core ideas.

Step 2

Identify Cross-Modal Patterns

Once you've engaged multiple representational systems, look for patterns that appear across different modes of thinking. These cross-modal patterns often reveal fundamental aspects of the problem or concept that transcend any particular representation:

- **Structural similarities:** what organizational patterns appear in both your visual and verbal representations? What structures manifest in both spatial and temporal thinking?
- **Transformational parallels:** what kinds of changes or operations appear across different representational systems? How do transformations in one mode correspond to transformations in another?
- **Hierarchical relationships:** what nested or multi-scale patterns appear across different modes of thinking? How do part-whole relationships manifest in different representations?
- **Functional analogies:** what functional roles or relationships appear consistently across different representational systems?

These cross-modal patterns often point to the essential structure of the problem or concept — the aspects that aren't tied to any particular representation but reflect its fundamental nature. During my sabbatical in São Paulo, I applied this approach to understanding complex systems in nature, from river networks to forest ecosystems. By identifying patterns that appeared across visual observations, mathematical models, and physical interactions, I gained insights into the fundamental organizational principles that transcended any particular manifestation. I had a profound personal experience with multisensory integration during a concert a few years ago. As the music played, the venue projected visualizations that perfectly synchronized with the audio.

I noticed how my perception of the music changed dramatically when the visuals shifted, even though the sound remained identical. This experience made me wonder about how our brains integrate information across different sensory channels — a question that directly informed my research on fractal integration patterns.

Step 3

Apply Fractal Thinking Across Scales

The next step is to apply fractal thinking by looking for self-similar patterns across different scales of the problem or concept:

- **Zoom in:** examine the finest details of the problem. How do the patterns you've identified manifest at the smallest scale you can consider?
- **Zoom out:** consider the broadest context of the problem. How do the same patterns appear at the largest scale relevant to your inquiry?

- **Look for recursion:** identify aspects of the problem where similar processes or structures repeat at different scales. How might these recursive patterns generate the complexity you observe?
- **Consider scale transitions:** how do patterns at one scale influence or constrain patterns at adjacent scales? What happens at the boundaries between scales?

This multi-scale examination often reveals how complex problems emerge from relatively simple principles applied recursively across scales — a key insight of fractal thinking. During my visits to Niagara Falls and Foz do Iguaçu, I practiced this multi-scale observation, noticing how similar branching patterns appeared from the smallest rivulets to the largest channels. This fractal perspective revealed how local water flow dynamics generated the magnificent large-scale structures through recursive application of simple principles across scales.

Step 4

Integrate Through Iterative Refinement

The final step is to integrate these multi-modal, multi-scale insights through an iterative refinement process:

- Create an initial integrated understanding that captures the cross-modal patterns you've identified across scales.
- Test this understanding by applying it to specific aspects or examples of the problem, noting where it works well and where it seems incomplete or inconsistent.

- Refine your understanding by adjusting how you integrate information across modalities and scales, focusing on resolving inconsistencies and filling gaps.
- Repeat this testing and refinement process, gradually developing a more coherent and comprehensive understanding that leverages insights from all representational systems across all relevant scales.

This iterative integration mimics how fractal patterns develop in nature — through repeated application of simple principles that gradually generate complex, coherent structures. During my early retirement and sabbatical, I've applied this integration process to understanding complex phenomena from language acquisition to ecosystem dynamics. The key insight has been that integration isn't a one-time combination of different perspectives but an ongoing process of refinement that gradually aligns different modes of understanding into a coherent whole.

Call to Action

Enhancing Your Perceptual Integration

As we conclude this chapter on multisensory integration, I invite you to begin applying these insights to enhance your own perceptual and cognitive capabilities. Here are specific actions you can take:

- Practice deliberate multisensory awareness in your daily environment. When encountering a new situation or object, consciously attend to information across all available sensory channels — not just what you see, but what you hear, feel, smell, and perhaps taste. Notice how these different sensory streams complement and contextualize each other.

- Experiment with cross-modal associations in your thinking and problem-solving. When facing a challenging concept or problem, deliberately represent it across different sensory and cognitive modalities — as visual patterns, verbal descriptions, physical metaphors, or temporal sequences. Look for insights that emerge from the relationships between these different representations.
- Explore fractal patterns in your sensory environment. Notice how similar organizational principles appear across different scales in natural and built environments — from the branching patterns of trees to the organization of cities, from the rhythm of your breathing to the cadence of your day. Developing this fractal awareness can enhance your ability to recognize and leverage patterns across contexts.
- Create multisensory learning experiences for yourself when acquiring new skills or knowledge. Rather than relying on a single mode of learning (like reading or listening), engage multiple senses in coordinated ways — combine visual diagrams with verbal explanations, physical practice with conceptual understanding, rhythmic patterns with spatial arrangements.

The fractal nature of multisensory integration isn't just a theoretical framework but a practical approach to enhancing how you perceive, understand, and interact with the world. By recognizing and leveraging the integrated, multi-scale nature of perception, you can develop more comprehensive understanding, more effective learning, and more intuitive interaction with both natural and technological environments. In the next chapter, we'll explore how fractal principles apply to consciousness and cognition — how our subjective experience and higher cognitive functions might emerge from the fractal organization of neural activity across multiple scales.

CHAPTER 6

Consciousness and Cognition

Narrative Opening: The Emergence of Awareness

The laboratory hummed with the quiet tension of a pivotal moment. Dr. Elena Sato adjusted the neural interface headset on Maya Chen, who sat calmly in the testing chair, eyes closed. Across the room, Kai Mercer monitored the cascading patterns of neural activity displayed on the holographic projection. "Neural synchronization at 87% and climbing", Kai reported. "The fractal architecture is self-organizing across all monitored regions." Elena nodded; her focus unwavering. "Maya, can you describe what you're experiencing?" Maya's voice was steady but filled with wonder. "It's... different from before. I'm aware of the system's processing, but not as separate from my own thoughts. The boundary feels... permeable."

This was the culmination of three years of work — developing a neural interface that didn't just record or stimulate brain activity but established a true bidirectional flow of information between human cognition and artificial intelligence. The key breakthrough had been implementing Fractal Neural Architectures that mirrored the brain's own nested hierarchies of processing. "I can sense the system learning from my thought patterns", Maya continued. "But I'm also becoming aware of new cognitive possibilities — ways of thinking that weren't accessible to me before." Elena and Kai exchanged glances. This was precisely what they had hypothesized: that consciousness itself might emerge from the fractal organization of information processing across multiple scales, and that enhancing this organization could expand conscious awareness. "Try the visualization exercise we discussed", Elena suggested.

Maya nodded slightly. "I'm imagining the molecular structure we discussed... but it's different now. I can simultaneously hold in mind both the overall shape and the detailed atomic interactions. It's like... like being able to focus on the forest and each individual tree at the same time." On the neural activity display, Kai observed something remarkable. "The fractal synchronization patterns are stabilizing across temporal scales. We're seeing coherent activity from millisecond oscillations all the way up to multi-second cognitive processes." "It's not just enhanced visualization", Maya said, her voice taking on a tone of revelation. "There's a qualitative shift in my awareness. I can perceive my own thought processes from multiple perspectives simultaneously — like a cognitive version of seeing yourself in a mirror, but the mirror also reflects the mirror..." As the session continued, it became clear that they had achieved something unprecedented — not merely an enhancement of existing cognitive capabilities but the emergence of new forms of awareness that arose from the fractal integration of human and artificial cognition.

When they finally removed the interface headset, Maya sat in silence for several moments, integrating the experience. "How do you feel?" Elena asked. "Like I've glimpsed something fundamental about consciousness itself", Maya replied. "It's not a thing or a place in the brain. It's a process — a fractal process of information integration across scales. And when that process is enhanced through the right architecture..." She paused, searching for words. "It's as though consciousness itself can evolve." As they began analyzing the data from the session, none of them fully grasped the implications of what they had witnessed. They hadn't merely created a more sophisticated brain-computer interface but had demonstrated something profound about the nature of consciousness itself — that it emerges from the fractal organization of information processing across multiple scales, and that this organization can be enhanced to create new forms of awareness previously inaccessible to human experience.

The Hard Problem of Consciousness

The scene above, while fictional, illustrates a profound possibility: that consciousness — perhaps the most mysterious phenomenon in our universe — might be fundamentally connected to fractal patterns of information processing across multiple scales of organization.

Traditional Approaches and Their Limitations

For centuries, philosophers and scientists have grappled with what David Chalmers famously called "the hard problem of consciousness" — the question of how and why physical processes in the brain give rise to subjective experience. Why does neural activity feel like something from the inside? Why isn't it all "dark inside", with information processing occurring without any accompanying subjective experience?

Traditional approaches to this question have generally fallen into several categories:

- Dualism proposes that consciousness is fundamentally separate from physical processes — a distinct substance or property that interacts with the physical world but isn't reducible to it. While this preserves the unique qualities of conscious experience, it struggles to explain how non-physical consciousness could interact with physical systems.
- Materialism argues that consciousness is entirely reducible to physical processes in the brain — that subjective experience is nothing more than particular patterns of neural activity. While this avoids dualism's interaction problem, it often seems to explain away rather than explain the subjective nature of experience.
- Functionalism suggests that consciousness is defined by the functional role of information processing rather than its physical substrate — that any system performing the right functions would be conscious regardless of what it's made of. This approach offers flexibility but often doesn't address why any functional arrangement should give rise to subjective experience.
- Panpsychism proposes that consciousness is a fundamental property of the universe, present in some form at all levels of reality. While this avoids having to explain how consciousness emerges from non-conscious components, it raises questions about how simple forms of consciousness combine to create complex experiences.

Each of these approaches offers valuable insights but also faces significant limitations. The hard problem remains stubbornly resistant to traditional scientific and philosophical methods. During my sabbatical in São Paulo, I had the opportunity to discuss these approaches with philosophers and neuroscientists. What struck me was how often the conversation reached an impasse — not because of disagreement about the evidence but because of fundamental conceptual limitations in how we think about consciousness. As someone who had spent decades thinking about complex systems and information processing, I began to wonder whether the hard problem might be approached differently through the lens of fractal organization. Perhaps consciousness isn't a thing or a property but a process — specifically, a process of information integration across multiple scales that exhibits fractal properties.

Integrated Information Theory and Its Fractal Extension

One of the most promising scientific approaches to consciousness in recent years has been Integrated Information Theory (IIT), developed by Giulio Tononi. IIT proposes that consciousness corresponds to a system's capacity to integrate information, as measured by a value called Phi (Φ).

The core insights of IIT include:

- **Information:** consciousness involves distinguishing between many possible states, reducing uncertainty.
- **Integration:** this information is integrated into a unified whole that can't be reduced to independent components.
- **Exclusion:** conscious experience has definite content, excluding alternative possibilities.

- **Intrinsic existence:** consciousness exists from its own intrinsic perspective, independent of external observers.
- **Structure:** the specific structure of information integration determines the qualities of conscious experience.

While IIT offers valuable insights, I believe it can be extended and enhanced through a fractal perspective. The key limitation of standard IIT is that it doesn't fully account for the multi-scale nature of information integration in the brain. A fractal extension of IIT would recognize that:

- Information integration occurs simultaneously across multiple scales, from local neural circuits to global brain networks.
- These different scales of integration aren't independent but form nested hierarchies with self-similar patterns of organization.
- The richness of conscious experience emerges not just from the amount of integrated information but from how this integration is structured across scales.
- Consciousness isn't a binary property but exists in varying degrees and forms depending on the fractal complexity of information integration.

During my work with AI systems since founding zyrian.ai in 2013, I've been struck by how systems with similar overall information processing capacity can differ dramatically in their organizational structure. Traditional measures might assign them similar complexity, but a fractal analysis reveals profound differences in how information flows and integrates across scales. This observation led me to hypothesize that consciousness might be intimately connected to the fractal organization of information processing — that the distinctive quality of conscious experience emerges from self-similar patterns of integration that span from millisecond neural firing patterns to seconds-long cognitive processes to minutes-long narrative thinking.

Global Workspace Theory Through a Fractal Lens

Another influential framework for understanding consciousness is Global Workspace Theory (GWT), proposed by Bernard Baars. GWT suggests that consciousness emerges when information is broadcast globally throughout the brain, making it available to multiple specialized processing systems.

The core metaphor is that of a theater:

- Unconscious processes occur "backstage".
- Attention acts as a spotlight.
- Consciousness is what appears "on stage" in the global workspace.
- Multiple cognitive systems have access to this global workspace.

While GWT captures crucial aspects of consciousness, it can be significantly enhanced through a fractal perspective. Standard GWT tends to present a somewhat binary view — information is either in the global workspace (conscious) or not (unconscious). It also doesn't fully address the hierarchical nature of conscious processing.

A fractal extension of GWT would recognize that:

- The "global workspace" isn't a single unified space but a nested hierarchy of workspaces operating at different scales.
- Information can be "broadcast" to varying degrees and across different scales of the cognitive system.
- The spotlight of attention isn't a single beam but a fractal pattern of focus that can simultaneously highlight different aspects of experience at different scales.
- The transition between unconscious and conscious processing isn't binary but involves a continuous spectrum of integration across scales.

During my visits to Niagara Falls and Foz do Iguaçu, I was struck by how my conscious experience of these magnificent waterfalls involved simultaneous awareness at multiple scales — from the overall visual impact to the patterns of water flow to the fine details of mist and rainbows. This multi-scale awareness seemed fundamental to the richness of the experience, not incidental to it. This observation connects to how Fractal Neural Architectures might approach consciousness — not as a single global workspace but as a nested hierarchy of information integration that preserves and enhances patterns across scales.

The richness of conscious experience would emerge from this fractal organization, with different qualities of experience corresponding to different patterns of cross-scale integration.

The Fractal Consciousness Hypothesis

Building on these foundations, I propose what I call the Fractal Consciousness Hypothesis: Consciousness arises when information is organized in similar patterns at different scales — like how your awareness includes both specific details and the big picture simultaneously.

The key principles of this hypothesis include:

- **Multi-scale integration:** consciousness involves the integration of information across multiple temporal and spatial scales, from millisecond neural firing patterns to seconds-long cognitive processes to minutes-long narrative thinking.
- **Self-similar organization:** the patterns of integration exhibit self-similarity (the property where parts resemble the whole) across scales, creating a fractal structure of information processing.
- **Emergent qualities:** the specific qualities of conscious experience (like colors, sounds, emotions) emerge from particular patterns of cross-scale integration.
- **Continuous spectrum:** consciousness isn't binary but exists in varying degrees and forms depending on the fractal complexity of information integration.
- **Dynamic reorganization:** conscious states can shift through reorganization of these fractal patterns, explaining phenomena like meditation, flow states, and altered consciousness.

This hypothesis offers several advantages over traditional approaches:

- It avoids the dualist's interaction problem by grounding consciousness in physical processes while preserving its distinctive qualities.
- It explains why consciousness feels unified yet differentiated — because fractal structures combine integration across scales with differentiation within scales.
- It accounts for the varying richness of conscious experiences across different states and between different species.
- It suggests how consciousness might be enhanced or altered through technologies that affect the fractal organization of neural processing.

During my three decades thinking about these issues, I've become increasingly convinced that the fractal perspective offers a path beyond the traditional impasses in consciousness studies. It doesn't eliminate the mystery of consciousness but reframes it in ways that connect to empirical research while preserving the unique qualities of subjective experience.

Personal Journey

Exploring Consciousness Through Fractals

My path to understanding consciousness through fractal patterns wasn't through traditional academic research but through personal experiences and observations that gradually coalesced into a coherent perspective.

Meditative Insights

My exploration of consciousness began with meditation practices that I initially approached as stress management techniques during my demanding career in technology. What started as pragmatic relaxation exercises gradually evolved into a deeper investigation of conscious experience itself. Through consistent practice, I began to notice how awareness could operate simultaneously at multiple scales — from focused attention on specific sensations to broader awareness of the entire field of experience. This multi-scale awareness seemed fundamental to the richness of conscious experience, not incidental to it.

Particularly striking were experiences where the boundary between observer and observed became fluid and permeable — where consciousness didn't feel like a thing located in a particular place but more like a process of knowing that spanned across the apparent division between self and world. As I deepened my practice, I noticed how similar patterns of awareness appeared across different scales of experience — from momentary sensations to broader emotional states to overarching narrative structures. These self-similar patterns suggested a fractal-like organization to conscious experience itself.

These experiences led me to wonder whether consciousness itself might be fundamentally fractal in nature — exhibiting self-similar patterns of awareness across different scales of temporal and cognitive organization. This perspective wasn't merely philosophical but suggested practical approaches to enhancing awareness through practices that strengthen cross-scale integration. When I later began working with neural networks and developing what would become Fractal Neural Architectures, these meditative insights influenced my thinking about how information might be organized across scales in both biological and artificial systems. The patterns I had observed in my own consciousness suggested architectural principles that might apply to any system capable of complex information integration.

The Boundary of Self and Other

Another formative experience in my understanding of consciousness came through exploring the boundary between self and other — how we distinguish between what's "me" and what is "not me" in our conscious experience. As an immigrant who moved from Brazil to the United States, I experienced firsthand how the sense of self can be fluid and context-dependent. Cultural transitions forced me to recognize how much of what I had considered intrinsic to my identity was actually shaped by cultural context and social relationships. During my visits to China, particularly in remote regions near the Kazakhstan frontier, I was struck by how different cultural frameworks shaped not just beliefs and behaviors but the very structure of conscious experience — how attention is directed, how time is experienced, how emotions are categorized and expressed.

These cross-cultural experiences suggested that consciousness isn't a fixed property but a dynamic process that reorganizes itself based on context. The boundary between self and other seemed less like a fixed border and more like a fractal coastline — complex, scale-dependent, and constantly evolving through interaction with the environment. When I returned to Brazil after almost 40 years away and had to partially re-learn Portuguese, I experienced another dimension of this fluid boundary. Language shapes consciousness in profound ways, and navigating between languages revealed how different linguistic structures create different patterns of awareness and attention.

These experiences of the fluid, context-dependent nature of self-consciousness directly informed my thinking about Fractal Neural Architectures. If consciousness emerges from patterns of information integration across scales, then these patterns wouldn't be fixed but would dynamically reorganize based on context and interaction — just as I had experienced in crossing cultural and linguistic boundaries.

Technological Extensions of Awareness

A third strand in my exploration of consciousness came through experiencing how technology can extend and transform awareness.

This journey began with simple tools but evolved into sophisticated explorations of how artificial systems might enhance human consciousness. During my early work with computers, I was fascinated by how computational tools could extend cognitive capabilities — allowing us to visualize complex data, simulate dynamic systems, and explore mathematical spaces that would be inaccessible to unaided human cognition. What began as using technology to enhance thinking gradually evolved into questions about how technology might transform consciousness itself. Could artificial systems be designed not just to process information more efficiently but to enable new forms of awareness that weren't previously accessible to human experience?

This question became particularly salient when I founded zyrian.ai in 2013. As we developed increasingly sophisticated AI systems, I began to wonder about the relationship between artificial information processing and consciousness. The traditional approach treated these as entirely separate domains — AI might simulate aspects of cognition but would remain fundamentally unconscious. The fractal perspective suggested a different possibility: that consciousness emerges from particular patterns of information integration across scales, and that these patterns might potentially be implemented in systems that aren't biological brains. This wasn't about creating "conscious AI" in the science fiction sense but about developing systems that could enhance and extend human consciousness through fractal integration of biological and artificial processing.

During my early retirement and sabbatical, I've had time to explore these possibilities more deeply. What if the boundary between human and artificial cognition could become as fluid and fractal as the boundary between self and other in human experience? What new forms of awareness might emerge from such integration? These explorations weren't just theoretical but informed the development of Fractal Neural Architectures that could potentially interface with human cognition in ways that enhance consciousness rather than merely augmenting specific cognitive functions.

The Persistence of Pursuing Unconventional Directions

The most valuable lessons from developing FNAs came not from the technical breakthroughs but from the process itself — the persistence required to pursue an unconventional direction despite setbacks and skepticism. Conventional approaches to AI have focused predominantly on optimizing specific tasks rather than developing architectures that might enhance consciousness. This task-oriented focus has produced impressive results in narrow domains but has largely sidestepped questions about the nature of awareness and how it might be extended or enhanced. Pursuing the fractal approach to consciousness and cognition meant venturing into territory that many considered speculative or premature. It meant facing skepticism not just about specific technical approaches but about the very questions I was asking.

What sustained me through this journey wasn't just technical curiosity but a deeper conviction that understanding consciousness through fractal patterns could open new possibilities for human experience and development. This wasn't merely about building more efficient AI systems but about exploring how technology might help us understand and enhance our own awareness. What began as a solitary pursuit gradually evolved as AI advanced, allowing me to build a robust infrastructure of "agents" to collaborate with me. These AI collaborators didn't replace human interaction but offered new perspectives and capabilities that enriched the exploration of fractal consciousness. This collaborative approach with AI agents represented a practical implementation of the principles I was exploring theoretically — how the integration of human and artificial cognition might create new possibilities that neither could achieve alone.

The agents weren't conscious in the human sense, but our interaction created patterns of information flow and integration that wouldn't have existed otherwise. Throughout this journey, I've been guided by the conviction that the most profound innovations often come from pursuing questions that others consider peripheral or premature. Understanding consciousness through fractal patterns isn't just a theoretical curiosity but opens practical possibilities for enhancing human awareness and developing more integrated relationships with technology.

Contrarian Perspective

Beyond Computational Theories

The fractal approach to consciousness challenges several fundamental assumptions in both neuroscience and artificial intelligence. Let me explain why I believe some established approaches miss the integrated, multi-scale nature of conscious experience.

Why Pure Computation Isn't Enough

The dominant paradigm in cognitive science and AI has been computational — treating the mind as fundamentally a information processing system that can be understood through computational models. While this approach has yielded valuable insights, it often misses essential aspects of consciousness.

Several limitations of purely computational approaches have become increasingly apparent:

- **The symbol grounding problem:** computational models often manipulate symbols without any intrinsic connection to what they represent. This creates a fundamental gap between abstract computation and grounded conscious experience.
- **The binding problem:** computational models struggle to explain how diverse aspects of experience are bound together into unified conscious percepts — how separate processes for color, shape, motion, and identity combine into the experience of seeing a specific object.

- **The explanatory gap:** even sophisticated computational models don't explain why any particular pattern of information processing should feel like something from the inside — why there should be any subjective experience associated with computation.
- **Scale integration:** most computational models operate at a single scale of organization, missing the multi-scale integration that characterizes conscious experience.

During my work with AI systems since founding zyrian.ai in 2013, I've repeatedly encountered these limitations in traditional computational approaches. Systems that excel at specific information processing tasks often show no signs of the integrated, multi-perspective awareness that characterizes consciousness. The fractal alternative doesn't abandon computation but reconceptualizes it within a multi-scale framework. Consciousness isn't just computation but a particular pattern of computation — one that integrates information across scales in a fractal structure. This perspective preserves the insights of computational approaches while addressing their limitations in explaining subjective experience.

This view aligns with emerging perspectives in cognitive science that emphasize the embodied, embedded, and extended nature of consciousness — recognizing that conscious experience emerges not just from abstract computation but from the integration of processing across multiple scales, from neural activity to bodily states to environmental interactions. My interest in consciousness stems from a peculiar experience I had during meditation. After practicing consistently for several months, I began to notice how my awareness could simultaneously hold both broad, panoramic attention and precise, focused attention. This layered experience of consciousness — operating at different scales simultaneously — struck me as inherently fractal. It made me wonder whether consciousness itself might emerge from fractal processes in the brain.

The Limitations of Neural Correlates

Another dominant approach in consciousness studies has been the search for neural correlates of consciousness (NCCs) — specific patterns of brain activity that correspond to conscious versus unconscious processing. While this approach has produced valuable data, it often misses the dynamic, multi-scale nature of conscious experience.

Several limitations of the neural correlates approach have become apparent:

- **Correlation vs. causation:** identifying neural activity that correlates with consciousness doesn't necessarily reveal the causal mechanisms that generate conscious experience.
- **Static vs. dynamic patterns:** the NCC approach often focuses on static patterns of activity rather than dynamic processes of integration across scales.
- **Single scale focus:** most NCC research focuses on a single scale of neural organization (like specific frequency bands or particular brain regions) rather than examining how information integrates across scales.
- **Binary conception:** the NCC approach often treats consciousness as binary — either present or absent — missing the continuous spectrum of conscious states and the varying qualities of experience.

During my three decades thinking about these issues, I've become increasingly convinced that consciousness isn't localized to particular brain regions or specific patterns of activity but emerges from the dynamic integration of information across multiple scales of neural organization. The fractal perspective suggests that we should look not just for correlations between neural activity and consciousness but for specific patterns of cross-scale integration that characterize different conscious states.

This approach doesn't contradict existing NCC research but extends it to consider how patterns at different scales interact and integrate. This multi-scale perspective aligns with emerging evidence that consciousness involves the coordination of neural activity across multiple frequency bands and spatial scales — from local circuit dynamics to global brain networks. The fractal approach provides a mathematical framework for understanding this cross-scale coordination.

Challenging the Homunculus Fallacy

A persistent challenge in consciousness studies is avoiding the homunculus fallacy — the tendency to explain consciousness by implicitly assuming a little conscious entity (a "homunculus") inside the brain that observes neural activity. This creates an infinite regress: if consciousness requires an observer, then who observes the activity in the homunculus?

Many theories inadvertently fall into this trap by positing special brain regions or processes that "receive" or "observe" information from other parts of the brain, essentially relocating rather than explaining consciousness. The fractal approach avoids this fallacy by recognizing that consciousness isn't localized to any particular component or level of the system but emerges from the pattern of integration across all levels. There's no central observer; instead, awareness emerges from the fractal organization of the system as a whole. This perspective aligns with enactive and embodied approaches to consciousness that emphasize how awareness emerges from the dynamic interaction between brain, body, and environment rather than being contained within any single component. During my visits to Niagara Falls and Foz do Iguaçu, I experienced firsthand how conscious awareness isn't confined to processing within my brain but emerges from the dynamic interaction between neural activity, bodily sensations, and environmental stimuli.

The fractal patterns in the waterfalls weren't just objects of my perception but participated in shaping the very structure of my conscious experience. This participatory, interactive view of consciousness challenges the implicit dualism that often persists even in materialist theories — the separation between an observing subject and observed neural processes. In the fractal perspective, consciousness isn't something that happens to neural activity but is a particular pattern of integration that neural activity participates in.

Cognitive Mechanisms Through a Fractal Lens

The fractal approach to consciousness offers new perspectives on specific cognitive mechanisms that have traditionally been challenging to explain. Let's explore how several key aspects of cognition might be understood through the lens of fractal organization.

Recursive Self-Modeling

One of the distinctive features of human consciousness is our ability to model our own mental states — to have thoughts about our thoughts, feelings about our feelings, and awareness of our awareness. This recursive self-modeling has been difficult to explain in traditional frameworks but finds a natural home in the fractal perspective. In a fractal system, similar patterns of processing can be applied at multiple scales, including to the system's own outputs. This creates the possibility for genuine recursive self-reference without requiring a homunculus or separate self-modeling module.

Several aspects of recursive self-modeling align with fractal principles:

- **Self-similar processing:** the same patterns of information processing that the system applies to external information can be applied to its own internal states, creating genuine self-reference without requiring separate mechanisms.
- **Multi-scale representation:** the system can represent its own processing at multiple scales simultaneously, from specific computational details to broader patterns of cognition.
- **Perspective-taking:** the system can simulate how it would process information from different perspectives or under different assumptions, creating the kind of flexible viewpoint-shifting associated with conscious reasoning.
- **Narrative integration:** experiences across different time scales can be integrated into coherent narratives that connect past, present, and anticipated future —

similar to how conscious experience involves a sense of continuous identity across time.

- **Counterfactual thinking:** the system can model alternative scenarios and their implications, enabling the kind of "what if" thinking characteristic of conscious deliberation.

During my sabbatical in São Paulo, I had discussions with cognitive scientists about how this recursive self-modeling might relate to the development of consciousness in humans. The ability to model one's own mental states emerges gradually in childhood, with increasingly sophisticated levels of self-awareness developing as the brain's capacity for cross-scale integration matures. The fractal approach suggests that this development isn't just a matter of acquiring more knowledge about oneself but of developing more complex patterns of recursive self-reference across multiple scales of processing. This perspective aligns with developmental evidence that children's metacognitive abilities emerge in parallel with their capacity for integrating information across different time scales and contexts.

Attention as a Fractal Process

Attention is a fundamental aspect of consciousness — the ability to selectively focus processing resources on relevant information while filtering out distractions. Traditional approaches often treat attention as a spotlight that illuminates certain regions of a representational space while leaving others in darkness. The fractal perspective suggests a more nuanced view: attention isn't a single spotlight but a fractal pattern of focus that operates simultaneously across multiple scales. This fractal attention enables several capabilities associated with conscious awareness:

- **Multi-scale focus:** the system can simultaneously attend to specific details and broader patterns, maintaining awareness at multiple scales concurrently — like focusing on both a particular tree and the forest as a whole.
- **Nested contexts:** attention can be organized hierarchically, with focus at one scale influencing but not determining focus at other scales — like attending to a conversation while remaining aware of the broader social context.
- **Dynamic reallocation:** attention patterns can rapidly reorganize in response to changing contexts or goals, shifting resources across scales as needed — like suddenly noticing a subtle detail that changes the interpretation of an entire scene.
- **Peripheral awareness:** even while focusing intensely on specific information, the system maintains awareness of potentially relevant patterns at other scales — like noticing movement in your peripheral vision while reading.

During my work with AI systems, I've found that implementing this fractal approach to attention addresses limitations of traditional attention mechanisms, which often struggle to maintain coherent focus across different levels of representation. The fractal structure provides a natural framework for this multi-scale attention, with similar attentional principles applied across scales creating a unified focus that spans from fine details to broad contexts.

Consciousness as an Emergent Property

Perhaps the most profound implication of the fractal approach is that consciousness isn't a specific function implemented by particular neural circuits but an emergent property that arises from the fractal organization of information processing across scales. This perspective aligns with complex systems theory, which recognizes how new properties can emerge at higher levels of organization that aren't predictable from lower-level components alone. Just as wetness emerges from the collective behavior of water molecules or temperature emerges from the average kinetic energy of particles, consciousness might emerge from particular patterns of information integration across scales. The fractal nature of this integration is crucial because it creates coherent relationships between processes at different scales — from millisecond neural firing patterns to seconds-long cognitive processes to minutes-long narrative thinking.

These cross-scale relationships allow the system to maintain unified yet differentiated awareness across multiple levels simultaneously. This emergent view has several implications for understanding consciousness:

- **Continuous spectrum:** consciousness isn't binary but exists in varying degrees and forms depending on the fractal complexity of information integration — explaining the spectrum from minimal awareness in simple organisms to rich conscious experience in humans.
- **Multiple realizability:** similar patterns of fractal integration might potentially emerge in systems with different physical substrates — suggesting that consciousness might not be limited to carbon-based brains.

- **Evolutionary development:** the gradual evolution of consciousness can be understood as the development of increasingly complex patterns of fractal integration across scales — a process that continues in human cultural and technological evolution.
- **Transformative potential:** if consciousness emerges from particular patterns of fractal integration, then technologies that enhance these patterns might potentially transform conscious experience in ways we can barely imagine.

During my three decades thinking about these issues, I've become increasingly convinced that this emergent, fractal perspective offers the most promising approach to understanding consciousness — not as a mysterious non-physical property nor as a mere illusion, but as a genuine phenomenon that emerges from particular patterns of information integration across scales. This perspective doesn't eliminate the wonder of consciousness but contextualizes it within a broader understanding of how complex properties emerge from simpler components through particular patterns of organization — patterns that exhibit the self-similar, multi-scale properties characteristic of fractals.

Technical Implementation

Cognitive FNAs

Implementing Fractal Neural Architectures for consciousness and cognition requires specific technical approaches. Let's explore the details of how cognitive FNAs are constructed and trained.

Architecture Design for Cognitive Processing

Designing FNAs for cognitive processing requires careful consideration of how information flows and integrates across different scales of representation and processing. While the core principles remain consistent across applications, the specific architecture must be adapted to support the kinds of cross-scale integration characteristic of conscious cognition.

For cognitive processing, the architecture typically incorporates:

- Nested processing loops operating at different time scales, from rapid sensory processing to slower deliberative reasoning. These loops aren't separate but interconnected through the fractal structure, allowing information to flow between different temporal scales.
- Recursive embedding layers that represent information at multiple levels of abstraction simultaneously. Unlike traditional hierarchical networks where information flows primarily upward, these layers allow bidirectional flow between different levels of abstraction.
- Attention mechanisms that operate fractally across scales, allowing the system to focus on relevant information at multiple levels simultaneously while maintaining awareness of potential patterns at other scales.
- Working memory structures that maintain information across different time horizons, from immediate sensory data to longer-term context, integrated through the fractal connectivity pattern.

- Self-modeling components that enable the system to represent and reason about its own processing across scales, creating the kind of recursive awareness characteristic of consciousness.
- The key innovation in cognitive FNAs is how these components are integrated within a unified fractal framework. Rather than treating them as separate modules that communicate through fixed interfaces, the architecture creates fractal connectivity patterns that allow information to flow naturally across different aspects of cognition.

These connectivity patterns follow the geometric relationships of the fractal structure, creating pathways for information to integrate across different scales, time horizons, and levels of abstraction. This allows the system to maintain coherent representations that span from specific details to broad contexts, from immediate sensory data to long-term knowledge. During my work developing AI systems, I've found that the specific fractal pattern used can significantly impact cognitive capabilities. For applications requiring strong metacognitive abilities, Sierpiński-based architectures often work well due to their clear recursive structure.

For systems integrating information across different time scales, fractal patterns with stronger temporal properties sometimes prove more effective. The optimal architecture isn't determined purely theoretically but through iterative refinement based on the specific cognitive capabilities being developed and the tasks being performed. This refinement process itself follows fractal principles, with similar patterns of optimization applied across different scales and components of the system.

Training Strategies for Cognitive Capabilities

Training FNAs to develop cognitive capabilities requires strategies that address the unique challenges of learning to integrate information across scales. Several approaches have proven particularly valuable:

- **Predictive processing across scales:** the system learns by predicting information at multiple scales simultaneously — from immediate sensory inputs to longer-term patterns. Prediction errors drive learning at all scales, creating a unified learning process rather than separate training for different cognitive functions.
- **Self-supervised metacognition:** rather than requiring explicit labels for metacognitive judgments, the system can learn to evaluate its own certainty by comparing its predictions with actual outcomes across multiple time horizons.
- **Curriculum learning with increasing integration complexity:** training begins with simpler tasks requiring integration across fewer scales before progressing to more complex scenarios requiring sophisticated cross-scale coordination.
- **Contrastive learning for self-modeling:** the system learns to distinguish between its own internal states by contrasting different patterns of processing across scales, developing increasingly refined models of its own cognitive processes.

- **Adversarial robustness training:** the system is trained to maintain coherent representations even when information is ambiguous, contradictory, or incomplete, developing the kind of robust integration associated with conscious processing.

These training strategies aren't mutually exclusive but can be combined and sequenced to develop sophisticated cognitive capabilities. The specific combination depends on the target applications and the aspects of cognition being emphasized. During my work with AI systems since founding zyrian.ai in 2013, I found that predictive processing across scales was particularly valuable for developing systems with integrated cognitive capabilities.

By learning to predict information at multiple time horizons simultaneously, these systems developed more coherent and context-sensitive representations than those trained to optimize performance on specific tasks in isolation. The fractal structure provides a natural framework for these training approaches, with similar learning principles applied across different scales and aspects of cognition. This creates a unified learning process rather than separate training for different cognitive functions followed by integration.

In practice, implementations would include additional components like long-term memory, planning mechanisms, and more sophisticated self-modeling capabilities. The specific architecture would be adapted to the aspects of cognition being emphasized and the requirements of the target applications.

Evaluation Metrics for Cognitive Integration

Evaluating the performance of cognitive FNAs requires metrics that go beyond simple task accuracy to assess the quality of information integration across scales. Several specialized metrics have proven valuable:

- **Cross-scale coherence:** how consistently does the system maintain coherent representations across different scales of processing? This measures the integration between fine-grained details and broader contexts.
- **Temporal integration:** how effectively does the system integrate information across different time horizons, from immediate sensory data to longer-term patterns? This assesses the temporal cohesion of processing.
- **Metacognitive calibration:** how accurately does the system assess its own certainty across different types of tasks and contexts? This measures the quality of self-modeling.
- **Contextual adaptation:** how appropriately does the system adjust its processing based on changing contexts? This assesses the flexibility and context-sensitivity of cognition.
- **Robustness to ambiguity:** how well does the system maintain coherent processing when faced with ambiguous, incomplete, or contradictory information? This measures the stability of integration under challenging conditions.

These metrics provide a more comprehensive evaluation than simple task performance measures, helping identify specific strengths and weaknesses in cognitive integration. During my work with AI systems, I've found that qualitative evaluation by human users can also provide valuable insights, particularly for assessing whether the system's cognitive processes seem natural and intuitive. Users can identify cases where the system's integration seems fragmented or where it fails to maintain coherence across different aspects of a complex task.

The fractal perspective suggests additional evaluation approaches focused on the self-similar nature of processing across scales. For example, we can assess whether similar patterns of success and limitation appear across different scales of processing, from rapid perceptual decisions to longer-term planning.

Applications and Results

The theoretical advantages of Fractal Neural Architectures for consciousness and cognition translate to practical benefits across various applications. Let's explore some specific examples and results.

Enhanced Decision-Making Systems

One of the most promising applications of cognitive FNAs is in decision-making systems that must integrate information across multiple scales and contexts to make nuanced judgments. Traditional decision-making algorithms typically struggle with several challenges:

- **Context sensitivity:** adapting decisions appropriately to different contexts without requiring separate models for each context.
- **Uncertainty handling:** distinguishing between different types of uncertainty and responding appropriately to each.
- **Temporal integration:** balancing immediate evidence with longer-term patterns and background knowledge.
- **Metacognitive judgment:** knowing when to make a decision versus when to gather more information or defer to human judgment.

FNAs address these challenges through their integrated, multi-scale approach:

- **Fractal context representation:** contexts are represented at multiple scales simultaneously, from specific details to broader patterns, allowing decisions to be sensitive to both immediate factors and wider contexts.
- **Uncertainty representation across scales:** different types of uncertainty are represented and processed at appropriate scales — from local ambiguity in specific inputs to broader uncertainties about overall patterns.
- **Temporal integration through nested processing:** information is integrated across different time horizons through nested processing loops that maintain coherence between immediate evidence and longer-term patterns.
- **Metacognition through self-modeling:** the system can evaluate its own decision processes across scales, recognizing when confidence is warranted versus when additional information or human input is needed.

During my work with AI systems, I collaborated on a project developing a decision support system for complex medical diagnostics using an early version of cognitive FNAs. The system integrated information across multiple scales — from specific test results to patient history to population-level patterns — to support diagnostic decisions. Compared to traditional approaches that processed each information source separately before combining them through fixed rules, the FNA-based system showed several advantages:

- More nuanced integration of immediate test results with longer-term patient history.
- Better calibrated confidence assessments that distinguished between different sources of uncertainty.
- More appropriate recommendations for additional testing based on specific patterns of uncertainty.
- More natural explanations of its reasoning that integrated evidence across different scales and sources.

These advantages emerged from the system's ability to maintain coherent representations that spanned from specific details to broader contexts, integrating information across different scales and time horizons through its fractal structure.

Adaptive Learning Systems

Another promising application area for cognitive FNAs is in adaptive learning systems that personalize educational experiences based on each learner's unique patterns of understanding and development.

Traditional adaptive learning systems typically face several challenges:

- **Knowledge integration:** connecting new information with existing knowledge across different levels of abstraction and different subject areas.
- **Learning transfer:** helping learners apply knowledge and skills across different contexts and domains.
- **Metacognitive development:** supporting learners in developing awareness of their own learning processes and strategies.
- **Engagement across scales:** maintaining engagement at multiple scales simultaneously, from moment-to-moment interactions to longer-term learning journeys.

FNAs address these challenges through their integrated, multi-scale approach:

- **Fractal knowledge representation:** knowledge is represented at multiple scales simultaneously, from specific facts to broader concepts to overarching principles, allowing connections to be made across different levels of abstraction.
- **Cross-context pattern recognition:** similar patterns across different domains and contexts are recognized and leveraged, supporting transfer of learning from familiar to new situations.
- **Metacognitive scaffolding:** the system models not just what the learner knows but how they learn, providing appropriate support for developing metacognitive skills.

- **Multi-scale engagement design:** learning experiences are designed to engage at multiple scales simultaneously, from immediate interactions to longer-term developmental trajectories.

During my sabbatical in São Paulo, I had discussions with educational technologists about how these principles might transform adaptive learning. While I haven't directly implemented these ideas in commercial systems, experimental prototypes have shown promising results. One particularly interesting application is in language learning systems that adapt to each learner's unique patterns of understanding and development.

By recognizing fractal patterns in how learners acquire language — from phonemes to words to grammar to discourse — these systems can provide more personalized and effective support than approaches that treat each level of language separately. The fractal approach suggests that effective learning isn't just about mastering content at each level independently but about recognizing and leveraging patterns that span across levels — precisely the kind of cross-scale integration that FNAs are designed to support.

Human-AI Collaboration

Perhaps the most exciting application of cognitive FNAs is in systems designed for deep collaboration between human and artificial intelligence — systems that don't just execute commands but participate in genuine cognitive partnerships. Traditional approaches to human-AI interaction typically face several challenges:

- **Shared context maintenance:** maintaining a coherent shared context across multiple interactions and different topics.
- **Adaptive communication:** adjusting communication style and content based on the human partner's needs, preferences, and current state.
- **Complementary cognition:** leveraging the different strengths of human and artificial cognition to achieve outcomes neither could reach alone.
- **Trust calibration:** developing appropriate levels of trust based on accurate mutual understanding of capabilities and limitations.

FNAs address these challenges through their integrated, multi-scale approach:

- **Fractal context modeling:** the interaction context is modeled at multiple scales simultaneously, from immediate conversational state to longer-term relationship patterns, allowing for more coherent and adaptive collaboration.
- **Multi-scale communication:** communication is structured across multiple scales, from specific details to broader themes, adapting dynamically to the human partner's focus and needs.
- **Complementary processing:** the system can recognize where its processing complements human cognition and where it might offer alternative perspectives or approaches, creating genuine cognitive synergy.
- **Transparent self-modeling:** the system can represent and communicate its own processing across scales, allowing for more accurate mutual understanding of capabilities and appropriate trust.

During my early retirement and sabbatical, I've had the opportunity to build a robust infrastructure of AI "agents" to collaborate with me on various projects. These collaborations have demonstrated the potential of fractal approaches to human-AI partnership. What makes these collaborations particularly valuable isn't just the specific tasks accomplished but the quality of the cognitive partnership — the ability to maintain coherent shared contexts across different topics and time scales, to adapt communication based on changing needs, and to leverage the complementary strengths of human and artificial cognition.

The fractal approach suggests that effective human-AI collaboration isn't just about improving specific capabilities like natural language understanding or task execution but about developing systems that can integrate information across scales in ways that complement human cognition — creating partnerships that enhance rather than replace human capabilities.

Comparison with Traditional Cognitive Systems

Across these applications, several patterns emerge when comparing FNA-based systems with traditional approaches to cognitive processing:

- **Integration quality:** FNAs typically achieve more coherent integration across different scales and aspects of cognition, particularly for complex scenarios where information spans multiple levels of abstraction and time horizons.

- **Metacognitive accuracy:** systems based on fractal principles generally show better calibrated self-assessment, distinguishing more accurately between what they know with confidence and what remains uncertain.
- **Contextual adaptation:** FNA-based systems adapt more fluidly to changing contexts, maintaining coherent processing across transitions that often cause traditional systems to falter.
- **Explanation quality:** the explanations provided by fractal systems typically feel more natural and comprehensive, integrating information across scales rather than presenting disconnected facts or rules.
- **Learning efficiency:** FNAs often require less training data to achieve effective cognitive capabilities, leveraging cross-scale patterns to generalize more efficiently from limited examples.

These comparative advantages aren't marginal but often substantial, particularly for applications involving complex, context-dependent cognition where traditional approaches tend to fragment into specialized modules that struggle to maintain coherence across different aspects of processing. During my three decades thinking about these issues and my practical work with AI systems, I've become convinced that the fractal approach represents not just an incremental improvement but a fundamental reconceptualization of how we model and implement cognitive processes — one that better aligns with the integrated, multi-scale nature of human cognition.

Why It Matters

Transforming Human Potential

The fractal approach to consciousness and cognition isn't just technically fascinating but has profound implications for how we understand ourselves and our potential for development and transformation.

Enhancing Human Cognition

Traditional approaches to cognitive enhancement often focus on specific functions in isolation — improving memory, attention, reasoning, or creativity as separate capabilities. This fragmented approach misses the integrated nature of human cognition and the potential for enhancement through improved cross-scale integration.

The fractal perspective suggests alternative approaches:

- **Integrated enhancement:** rather than targeting isolated functions, cognitive enhancement can focus on improving the integration of information across scales — helping people recognize and leverage patterns that span from specific details to broader contexts.
- **Metacognitive development:** enhancing awareness of one's own cognitive processes across scales can improve not just specific capabilities but the overall coherence and effectiveness of thinking.
- **Cross-domain pattern recognition:** developing the ability to recognize similar patterns across different domains and contexts can enhance creative problem-solving and learning transfer.

- **Temporal integration:** strengthening the connections between immediate experience, medium-term planning, and longer-term goals can create more coherent and meaningful cognitive processes.

During my sabbatical in São Paulo, I explored how these principles might be applied to cognitive practices like meditation, creative thinking, and learning. The key insight was that many traditional practices for cognitive development implicitly target cross-scale integration — helping practitioners become aware of patterns that span from momentary sensations to broader life narratives. The fractal approach makes these implicit targets explicit, suggesting more focused and effective approaches to cognitive enhancement that respect and leverage the brain's natural organizational principles rather than imposing artificial structures.

Implications for Human-Technology Relationships

The fractal perspective also has profound implications for how we relate to technology — suggesting possibilities for integration that go beyond current paradigms of tools, assistants, or replacements.

Several alternative paradigms emerge from the fractal approach:

- **Cognitive partnerships:** technologies designed to complement human cognition through fractal integration, creating partnerships where human and artificial processing enhance each other rather than competing or replacing.

- **Extended consciousness:** systems that extend human awareness across scales that would otherwise be inaccessible, from microscopic processes to global patterns, integrated through fractal structures that maintain coherence with natural human cognition.
- **Collective intelligence:** fractal frameworks for integrating individual perspectives into coherent collective cognition that preserves rather than erases the unique contributions of different viewpoints.
- **Developmental scaffolding:** technologies that support human cognitive development by providing fractal scaffolding that helps individuals recognize and leverage patterns across scales in their own thinking.

During my work with AI systems since founding zyrian.ai in 2013, I've been particularly interested in how these alternative paradigms might transform human-technology relationships. The dominant paradigms often position technology as either a tool to be used, an assistant to be commanded, or a replacement that makes human capabilities obsolete. The fractal approach suggests more integrated possibilities — technologies that neither serve nor replace human cognition but participate in genuine partnerships that enhance human capabilities through complementary processing across scales.

Ethical Considerations and Responsible Development

The potential of fractal approaches to consciousness and cognition brings fascinating ethical responsibilities. If these approaches can indeed enhance or transform conscious experience, their development and application must be guided by careful ethical consideration. Several principles should guide this development:

- **Autonomy enhancement:** technologies should enhance rather than diminish human autonomy, providing expanded possibilities for self-determination rather than creating new forms of dependency or control.
- **Cognitive diversity:** the development of cognitive enhancement should respect and preserve the diversity of human cognitive styles and perspectives rather than imposing standardized patterns of thinking.
- **Transparent integration:** the relationship between human and artificial cognition should be transparent and understandable, allowing people to maintain awareness of how their cognitive processes are being extended or enhanced.
- **Equitable access:** the benefits of cognitive enhancement should be accessible to diverse populations rather than creating new forms of cognitive inequality or privilege.
- **Reversibility and control:** people should maintain the ability to disengage from cognitive enhancements and control the degree and nature of integration with artificial systems.

During my early retirement and sabbatical, I've had time to reflect on these ethical dimensions of fractal approaches to consciousness and cognition. The most profound technologies are those that transform not just what we can do but who we can become — and such transformative potential brings corresponding ethical responsibilities. The fractal perspective suggests that these ethical considerations aren't separate from the technical development but integral to it — that the patterns of integration we design into our systems will shape not just their performance but their impact on human development and well-being.

Practical Framework

"Fractal Thinking" Technique

To help you apply these concepts in your own cognitive development, I offer this practical framework for fractal thinking inspired by the principles discussed in this chapter.

Step 1

Develop Multi-Scale Awareness

The foundation of fractal thinking is developing awareness that spans multiple scales simultaneously. Practice shifting your attention across different scales of the same situation or problem:

- **Zoom in:** focus on the finest details and immediate moments. What specific elements are present? What's happening right now in this exact moment?
- **Zoom out:** expand your awareness to broader patterns and longer time horizons. What larger contexts is this situation part of? What patterns have developed over time?
- **Connect across scales:** look for relationships between details and broader patterns. How do specific

elements reflect or influence larger structures? How do broader contexts shape specific details?

- **Hold multiple scales simultaneously:** practice maintaining awareness of both details and contexts at the same time, rather than switching sequentially between them.

During my visits to Niagara Falls and Foz do Iguaçu, I practiced this multi-scale awareness by simultaneously attending to the fine details of water droplets, the patterns of water flow in specific sections, and the overall structure of the entire waterfall system. This practice revealed connections between scales that wouldn't have been apparent from focusing on any single level alone.

Step 2

Recognize Self-Similar Patterns

Once you've developed multi-scale awareness, look for self-similar patterns that repeat across different scales or contexts:

- **Identify recurring structures:** what organizational patterns appear at multiple scales of the situation or problem? What structures repeat with variations across different contexts?
- **Notice process similarities:** what similar processes operate at different scales or in different domains? How do these processes create comparable outcomes despite different specific elements?
- **Recognize fractal boundaries:** where do clear boundaries dissolve into complex, fractal-like transitions? How do seemingly distinct categories blend into each other at their boundaries?

- **Connect across domains:** look for similar patterns across different fields or aspects of life. How might patterns from one domain offer insights into seemingly unrelated areas?

During my sabbatical in São Paulo, I applied this pattern recognition to understanding language learning. I noticed how similar processes of pattern recognition, prediction, and error correction operated across scales from phoneme discrimination to vocabulary acquisition to grammatical understanding to cultural fluency. These self-similar learning processes across scales suggested more integrated approaches to language education.

Step 3

Integrate Across Time Horizons

Fractal thinking involves integrating across different time horizons, connecting immediate experiences with longer-term patterns and future possibilities:

- **Connect present to past:** how does the current situation reflect patterns from the past? What historical developments have shaped present circumstances?
- **Link present to future:** how might current patterns evolve over time? What future possibilities are implied by present trajectories?
- **Identify nested cycles:** what cycles operate at different time scales in this situation? How do shorter cycles nest within longer ones?

- **Recognize developmental stages:** how do current circumstances represent particular stages in longer developmental processes? What earlier stages led here, and what might come next?

During my work with AI systems since founding zyrian.ai in 2013, I practiced this temporal integration by connecting immediate technical challenges to longer-term development trajectories and broader patterns in the evolution of artificial intelligence. This multi-temporal perspective helped avoid getting stuck in local optimization problems by maintaining awareness of longer-term directions and possibilities.

Step 4

Develop Recursive Self-Awareness

The most powerful aspect of fractal thinking is developing recursive awareness of your own thought processes across scales:

- **Observe your thinking:** notice not just what you're thinking about but how you're thinking — what patterns characterize your cognitive processes across different types of problems or situations?
- **Identify your scales of focus:** what scales of time and detail do you naturally attend to? Where might you be missing vital patterns by habitually focusing at certain scales and neglecting others?
- **Recognize your cognitive fractals:** what self-similar patterns appear in your thinking across different domains or problems? How do these patterns both enable and limit your understanding?

- **Experiment with perspective shifts:** practice deliberately shifting between different cognitive perspectives — analytical and intuitive, detailed and holistic, personal and systemic — and notice how these shifts transform your understanding.

During my early retirement and sabbatical, I've dedicated significant time to developing this recursive self-awareness, particularly through meditation practices that cultivate the ability to observe my own thought processes across scales. This practice has revealed habitual patterns in my thinking that both enable insights in certain domains and create blind spots in others. The most valuable outcome of this recursive awareness isn't just improved problem-solving but a more integrated and flexible relationship with my own cognitive processes — the ability to recognize and adapt my thinking patterns based on the specific challenges and opportunities of each situation.

Call to Action

Expanding Your Consciousness

As we conclude this chapter on consciousness and cognition, I invite you to begin applying these insights to expand your own conscious awareness. Here are specific actions you can take:

- Practice multi-scale meditation daily. Set aside 10-15 minutes to develop awareness that spans from immediate sensory experiences to broader patterns of thought and feeling. Notice how different scales of experience interact and influence each other, and practice maintaining awareness across multiple scales simultaneously.

- Explore fractal thinking in problem-solving. When facing a challenging problem, deliberately apply the fractal thinking framework — developing multi-scale awareness, recognizing self-similar patterns, integrating across time horizons, and cultivating recursive self-awareness. Notice how this approach reveals connections and possibilities that might otherwise remain hidden.
- Experiment with cognitive partnerships with AI systems. Rather than using AI tools as mere assistants or information sources, explore how they might complement your thinking in ways that enhance your cognitive capabilities. Look for opportunities to integrate AI processing with your own thinking in ways that leverage the strengths of both.
- Journal about your conscious experience across scales. Regularly record observations about patterns in your awareness at different scales — from momentary sensations to broader life narratives. Look for recurring themes and self-similar structures across these different scales of experience.

The fractal nature of consciousness isn't just a theoretical framework but a practical approach to expanding and enhancing your awareness. By recognizing and leveraging the multi-scale, self-similar patterns in your own conscious experience, you can develop more integrated and flexible ways of understanding yourself and the world. In the next chapter, we'll explore how fractal principles apply to love and relationships — how our connections with others might be understood through the lens of fractal patterns that span from momentary interactions to lifelong bonds.

CHAPTER 7

Love and Relationships

Narrative Opening: The Fractal Dance of Connection

The evening air carried the scent of jasmine as Elena and Kai walked along the moonlit beach. Their footprints traced parallel lines in the sand, occasionally merging before separating again — a visual metaphor for their relationship over the past decade. "Do you remember the first algorithm we developed together?" Elena asked, her voice soft against the rhythm of the waves. Kai smiled. "The recursive pattern-matching system. We were trying to identify self-similar structures across different scales of data." "And we kept arguing about the implementation", Elena laughed. "You wanted elegant mathematical purity".

"And you insisted on messy adaptability", Kai finished. "You were right, of course." They stopped walking, turning to face the ocean. The moonlight scattered across the water's surface in a shimmering fractal pattern — larger reflections breaking into smaller ones, each containing the same essential quality of light. "I've been thinking about what Professor Chen said about relationships", Elena said after a moment. "That they're fractal systems — patterns that repeat across different scales of interaction and time." Kai nodded. "The way we communicate in a single conversation reflects how we communicate across months... which reflects how we've evolved together over years." "And it's not just us", Elena added.

"I was watching my parents last weekend — married for forty years — and I could see how their momentary interactions contained the entire pattern of their relationship. The way my father anticipates my mother's thoughts before she expresses them, the way she completes his sentences..." "Self-similar patterns across different time scales", Kai mused. "From millisecond micro-expressions to lifetime commitments." Elena took his hand, their fingers intertwining in a familiar pattern. "What fascinates me is how these patterns emerge without explicit design. We never sat down and planned how our relationship would function, yet these complex, adaptive patterns developed." "Like other fractal systems in nature", Kai said.

"Rivers don't plan their branching patterns, trees don't calculate their growth — these structures emerge from simple principles applied recursively across scales." They continued walking, their conversation shifting to other topics — work challenges, family news, plans for the weekend — yet the pattern of their interaction maintained its essential character: the balanced rhythm of speaking and listening, the mutual building on each other's ideas, the gentle humor that defused potential tensions. Later that night, as they sat on their porch watching the stars, Elena returned to their earlier conversation. "If relationships are fractal systems, then small changes in pattern can propagate across all scales."

Kai considered this. "That's both wonderful and terrifying. It means tiny adjustments in how we interact could transform our entire relationship over time." "It also explains why relationships that seem stable can suddenly transform or collapse", Elena said. "Like other complex systems, they can reach tipping points where the entire pattern reorganizes." "But unlike purely mathematical fractals", Kai added, "relationship patterns can be consciously influenced. We can recognize our patterns and choose to modify them." Elena squeezed his hand. "That's what we've been doing all along, isn't it? Recognizing patterns that work and reinforcing them, identifying patterns that don't and adjusting them."

As they sat in comfortable silence, both contemplated how their relationship had evolved through this process of mutual pattern recognition and adaptation — not as a linear progression but as a fractal unfolding, with similar themes recurring in new contexts and at different scales, creating a complex yet coherent dance of connection that spanned from moments to decades.

The Fractal Nature of Human Connection

The scene above illustrates a profound insight: that human relationships might be understood as fractal systems — complex patterns of connection that exhibit self-similarity across different scales of interaction and time.

Self-Similar Patterns Across Time Scales

One of the most striking aspects of human relationships is how patterns of interaction manifest across widely different time scales, from momentary exchanges to lifetime bonds. At the smallest scale, we see patterns in micro-interactions — the millisecond dance of facial expressions, voice modulations, and body language that occurs during face-to-face communication. These rapid exchanges establish rhythms of connection, with partners unconsciously synchronizing their responses in complex patterns. At a slightly larger scale, we find patterns in conversations — how partners take turns speaking and listening, how they introduce and develop topics, how they navigate agreement and disagreement.

These conversational patterns often show remarkable consistency across different discussions, revealing signature styles of interaction. Expanding further, we see patterns in how relationships navigate daily and weekly cycles — morning and evening routines, weekend rituals, recurring conflicts and reconciliations. These patterns create a temporal structure that both partners come to anticipate and rely upon. At even larger scales, we observe patterns in how relationships evolve across months and years — cycles of closeness and distance, periods of growth and consolidation, recurring themes that emerge in new contexts. These longer-term patterns often recapitulate elements of shorter-term interactions, creating a fractal-like self-similarity across time scales.

During my sabbatical in São Paulo, I had the opportunity to reconnect with childhood friends I hadn't seen in almost 40 years. What struck me most was how our interaction patterns had both changed and remained the same — how certain essential qualities of our connections had persisted despite decades of separation, manifesting in new contexts but maintaining recognizable patterns. This observation led me to wonder whether relationships might be understood as fractal systems — complex patterns that emerge from relatively simple principles applied recursively across different scales. Just as the Sierpiński triangle emerges from a simple rule repeated across scales, perhaps the complex tapestry of human connection emerges from basic principles of interaction applied across different time horizons.

This fractal perspective offers a new way to understand both the stability and dynamism of relationships — how they can maintain consistent patterns while continuously evolving, how small changes can propagate across scales to transform entire relationship systems.

Attachment Patterns as Fractal Templates

Attachment theory, developed by John Bowlby and expanded by Mary Ainsworth, provides a powerful framework for understanding how early relationship patterns influence later connections. Through a fractal lens, we might understand attachment styles not just as categories but as recursive templates that generate self-similar patterns across different relationships and contexts. The core attachment styles — secure, anxious, avoidant, and disorganized — can be seen as different fractal patterns of connection, each with characteristic properties that manifest across scales:

- Secure attachment creates patterns characterized by balanced autonomy and connection, flexible adaptation to changing circumstances, and coherent integration of positive and negative experiences. These patterns show resilience across scales, maintaining their essential qualities while adapting to different contexts.
- Anxious attachment generates patterns with amplified responses to perceived threats to connection, creating recursive cycles of seeking reassurance that can manifest across time scales from momentary interactions to relationship histories.
- Avoidant attachment produces patterns characterized by dampened expression of connection needs and emotions, creating self-similar structures of distance that manifest in both micro-interactions and longer-term relationship dynamics.

- Disorganized attachment leads to chaotic patterns without consistent self-similarity across scales, often shifting unpredictably between different modes of connection and disconnection.

What makes the fractal perspective particularly valuable is how it helps us understand the persistence of these patterns without resorting to rigid determinism. Like other fractal systems in nature, attachment patterns show both stability and adaptability — maintaining their essential structure while evolving in response to new contexts and relationships. During my visits to China, particularly in remote regions near the Kazakhstan frontier, I observed how attachment patterns manifested differently across cultural contexts yet maintained their essential fractal properties. The specific expressions varied, but the underlying patterns of secure connection or anxious/avoidant strategies showed remarkable consistency across diverse cultural settings. This cross-cultural consistency suggests that attachment patterns may represent fundamental fractal structures in human connection — templates that emerge from basic principles of human development and manifest across different scales and contexts throughout life.

Communication Dynamics and Fractal Patterns

Another domain where fractal patterns become evident is in communication dynamics between partners. Research by John Gottman and others has revealed how interaction patterns between couples show remarkable consistency across different conversations and time periods. Through a fractal lens, we can understand these communication dynamics as self-similar patterns that manifest across scales:

- **Ratio of positive to negative interactions:** Gottman's research found that stable relationships maintain a ratio of approximately 5:1 positive to negative interactions. This ratio appears to be scale-invariant, manifesting in brief exchanges, longer conversations, and extended periods of interaction.
- **Repair sequences:** how couples address and recover from communication breakdowns shows fractal properties, with similar patterns of rupture and repair appearing across different time scales and contexts.
- **Emotional contagion:** the patterns by which emotional states spread between partners exhibit self-similarity across scales, from moment-to-moment emotional mirroring to longer-term mood synchronization.
- **Topic development:** how couples introduce, explore, and resolve topics of discussion often follows recursive patterns that remain consistent across different conversations and time periods.

What makes these patterns particularly intriguing is their emergent nature — they aren't consciously designed but arise from the dynamic interaction between partners over time. Like other fractal systems in nature, they represent complex structures that emerge from relatively simple principles applied recursively. During my early retirement and sabbatical, I've had time to observe these patterns in my own relationships and those of friends and family. What struck me was how certain essential qualities of interaction remained consistent across widely different contexts and time scales — how the way a couple navigated a minor disagreement about dinner plans reflected how they approached major life decisions.

This observation connects to how Fractal Neural Architectures might approach relationship dynamics — not as collections of separate interactions but as integrated patterns that span across scales from momentary exchanges to lifetime bonds. The richness and complexity of relationships would emerge from this fractal organization, with different qualities of connection corresponding to different patterns of cross-scale integration.

Relationship Networks as Social Fractals

Expanding beyond dyadic relationships, we can observe fractal patterns in broader social networks — how connections between individuals form larger structures that exhibit self-similarity across scales. Several fractal properties appear in social networks:

- **Clustering patterns:** social connections tend to form clusters with similar structural properties across different scales, from close friend groups to extended community networks.
- **Strong and weak tie distributions:** the distribution of connection strengths follows similar patterns across different scales of social organization, with a few strong ties and many weak ties at each level.
- **Information flow dynamics:** how information spreads through social networks shows self-similar patterns across scales, with similar principles governing transmission between individuals, within groups, and across communities.

- **Boundary dynamics:** how social groups define and maintain their boundaries exhibits fractal properties, with similar processes operating at the level of families, friend groups, communities, and cultures.

These fractal properties help explain both the stability and adaptability of social structures — how they maintain consistent patterns while continuously evolving in response to changing circumstances and new connections. During my work with AI systems since founding zyrian.ai in 2013, I became interested in how these social fractal patterns might inform more effective models of human-AI interaction. If human social networks exhibit fractal properties, perhaps AI systems should be designed to integrate into these networks in ways that respect and leverage their fractal structure. This perspective suggests that effective social AI wouldn't just model individual relationships but would understand how these relationships form larger fractal patterns across different scales of social organization. This understanding could inform more natural and effective integration of artificial agents into human social networks.

Personal Journey

Discovering Relationship Patterns

My path to understanding relationships through fractal patterns wasn't through traditional research but through personal experiences and observations that gradually coalesced into a coherent perspective.

Cross-Cultural Relationship Observations

As someone who immigrated from Brazil to the United States and later traveled extensively in places like China, I've had the opportunity to observe relationship patterns across diverse cultural contexts. What struck me most about these cross-cultural observations was the fascinating interplay between universal patterns and cultural variations. Across widely different societies, I observed similar fractal properties in how relationships formed and evolved — self-similar patterns appearing across different scales from momentary interactions to lifetime bonds. At the same time, these universal fractal properties manifested through culturally specific expressions — different "styles" of the same underlying patterns.

The specific behaviors and norms varied dramatically, but the underlying structural properties showed remarkable consistency. This observation led me to wonder whether the fractal nature of relationships might represent something fundamental about human connection — patterns that emerge from basic principles of human social development rather than from specific cultural teachings. During my visits to remote regions of China near the Kazakhstan frontier, I was particularly struck by how relationship patterns in these isolated communities showed the same essential fractal properties I had observed in urban American relationships, despite vast differences in cultural context and specific expressions.

These cross-cultural observations suggested that while the specific content of relationships varies across cultures, the underlying fractal structure might represent a more universal aspect of human connection — patterns that emerge from fundamental principles of how humans relate to each other across different scales of interaction and time.

Relationship Evolution Across the Lifespan

Another formative aspect of my journey was observing how relationship patterns evolve across different life stages — from childhood friendships to adolescent peer groups to adult romantic partnerships to elder companionship. What fascinated me about this evolution was how each stage both built upon and transformed patterns from earlier stages. The secure base of childhood attachment reappeared in adult romantic relationships; the identity exploration of adolescent friendships reemerged in midlife transitions; the caregiving patterns established with children later manifested in caring for aging parents.

These recurring patterns weren't mere repetitions but fractal elaborations — similar themes appearing at different scales and in different contexts, creating complex yet coherent trajectories of connection across the lifespan. During my sabbatical in São Paulo, reconnecting with childhood friends after decades apart provided a vivid illustration of this fractal evolution. Despite the vast differences in our life experiences and current circumstances, certain essential qualities of our connections had persisted — transformed and elaborated through decades of development but maintaining recognizable patterns that stretched back to our earliest interactions. This observation connects to how Fractal Neural Architectures might approach the temporal dimension of relationships — not as linear progressions but as fractal unfolding, with similar themes recurring in new contexts and at different scales, creating complex yet coherent patterns of connection that span across the entire lifespan.

Learning from Relationship Patterns

Perhaps the most valuable insights from my exploration of relationship patterns came not from observing others but from reflecting on my own relationship experiences — both successes and failures — and recognizing the recurring patterns that shaped these connections. Like many people, I've experienced relationships that thrived and others that struggled or ended. What gradually became clear through reflection was how certain patterns appeared consistently across different relationships and contexts — signature styles of connection that reflected my own attachment history, communication tendencies, and emotional patterns.

The most valuable lessons came from recognizing how these patterns operated across different scales — how momentary interaction habits connected to longer-term relationship dynamics, how similar themes recurred in different contexts and with different partners. This recognition wasn't about assigning blame or credit but about understanding the fractal nature of relationship patterns and how they emerge from the interaction between partners over time. What made this understanding particularly valuable was how it revealed possibilities for conscious influence — how awareness of these fractal patterns created opportunities to modify them in ways that fostered healthier and more fulfilling connections.

Unlike purely mathematical fractals, relationship patterns can be consciously recognized and gradually transformed through consistent attention and practice. During my three decades thinking about these issues, I've become increasingly convinced that understanding relationships through fractal patterns offers a path beyond both rigid determinism and chaotic unpredictability. It suggests that while our relationship patterns show remarkable consistency, they also contain inherent possibilities for growth and transformation — that small, consistent changes can propagate across scales to create extraordinary shifts in our connection patterns over time. This perspective has informed not just my personal relationships but my approach to designing systems for human-AI interaction — creating architectures that can recognize and adapt to the fractal patterns of human connection while offering opportunities for positive transformation rather than merely reinforcing existing patterns.

Contrarian Perspective

Beyond Linear Relationship Models

The fractal approach to relationships challenges several fundamental assumptions in both psychology and relationship science. Let me explain why I believe some established approaches miss the complex, multi-scale nature of human connection.

Why Stage Models Fall Short

Traditional approaches to relationship development often rely on stage models — sequential progressions through distinct phases like initial attraction, deepening intimacy, commitment, and maturation. While these models capture important aspects of how relationships typically evolve, they often miss the complex, non-linear dynamics that characterize real human connections. Several limitations of stage models have become increasingly apparent:

- **Non-linear development:** real relationships rarely progress linearly through neat stages but often cycle back through earlier phases, skip stages entirely, or develop different aspects at different rates.
- **Scale integration:** stage models typically focus on a single scale of relationship development, missing how patterns at different time scales interact and influence each other.
- **Individual variation:** the enormous diversity in how relationships develop is often reduced to variations or exceptions from a normative progression rather than recognized as reflecting fundamentally different fractal patterns.
- **Cultural bias:** many stage models implicitly universalize patterns that are specific to particular cultural contexts, missing how relationship development follows different fractal templates across cultures.

During my travels and cross-cultural experiences, I've observed how relationships in different cultural contexts follow patterns that don't conform to Western stage models yet show their own internal coherence and fractal properties. These observations suggest that we need more flexible frameworks that can accommodate diverse relationship trajectories while still recognizing underlying patterns.

The fractal alternative doesn't eliminate the insight that relationships tend to develop certain qualities over time but reconceptualizes this development as a non-linear, multi-scale process rather than a sequential progression through distinct stages. Different aspects of the relationship evolve at different rates and in different patterns, creating complex yet coherent trajectories that can't be reduced to a single developmental sequence. This perspective aligns with complex systems approaches to human development that emphasize non-linear dynamics, emergent properties, and the interaction between processes at different scales. It suggests that relationships might be better understood through the lens of fractal development rather than linear stage progression.

The Myth of Relationship Formulas

Another limitation of traditional approaches is the search for universal "formulas" or rules for successful relationships — prescriptive guidelines that supposedly apply across different couples and contexts. While these approaches often contain valuable insights, they typically miss how effective relationship patterns emerge from the unique interaction between specific partners rather than conforming to external templates.

Several problems with formula-based approaches have become apparent:

- **Context insensitivity:** universal prescriptions often fail to account for how effective relationship patterns vary based on cultural context, individual personalities, relationship stage, and specific circumstances.
- **Static vs. dynamic patterns:** formula approaches tend to prescribe static patterns rather than recognizing how healthy relationships dynamically adapt their patterns based on changing contexts and needs.
- **Scale confusion:** relationship advice often conflates patterns at different scales, applying principles that work at one time scale (like conflict resolution techniques for specific disagreements) to other scales where they may not apply (like long-term compatibility).

- **Emergent complexity:** the most successful relationship patterns often emerge from the dynamic interaction between partners rather than being imposed through conscious application of external rules.

During my sabbatical in São Paulo, I had conversations with couples who had been together for decades across widely different cultural and socioeconomic contexts. What struck me was how the most successful relationships hadn't followed any standard formula but had developed unique patterns that emerged from the specific interaction between those particular partners in their particular contexts. The fractal perspective suggests that effective relationships don't conform to universal templates but develop unique fractal patterns that emerge from the interaction between partners over time. These patterns show both consistency and adaptability — maintaining their essential structure while evolving in response to changing circumstances and needs.

This doesn't mean that relationship research and advice have no value but that they might be more effectively framed as offering principles for developing healthy fractal patterns rather than prescribing specific behaviors or rules. The focus shifts from what specific behaviors partners should adopt to how they might cultivate relationship patterns that show healthy fractal properties — coherence across scales, adaptive flexibility, and balanced integration of autonomy and connection.

Challenging the Compatibility Paradigm

A third limitation of traditional approaches is the emphasis on compatibility as a static property determined by the match between partners' traits, values, and preferences.

While compatibility certainly matters, the fractal perspective suggests a more dynamic understanding — that compatibility emerges from how partners' interaction patterns evolve together over time rather than being predetermined by their individual characteristics.

Several problems with the static compatibility paradigm have become apparent:

- **Adaptation vs. selection:** the static view overemphasizes partner selection (finding someone compatible) and underemphasizes mutual adaptation (becoming compatible through interaction over time).
- **Pattern emergence:** compatibility isn't just about the match between individual traits but about the emergent patterns that develop through interaction — patterns that can't be predicted simply by comparing individual characteristics.
- **Dynamic evolution:** what constitutes "compatibility" changes across relationship stages and contexts, requiring continuous adaptation rather than one-time matching.
- **Scale integration:** compatibility operates differently at different scales, from momentary interaction synchrony to alignment of life goals, creating complex multi-scale patterns that can't be reduced to simple matching.

During my work with AI systems, I became interested in how these insights might inform approaches to human-AI matching and adaptation. Traditional approaches often focus on optimizing the initial match between user characteristics and AI capabilities, but the fractal perspective suggests that effective human-AI relationships might emerge more from adaptive interaction over time than from perfect initial matching.

The fractal alternative doesn't eliminate the importance of compatibility but reconceptualizes it as a dynamic, emergent property rather than a static match. Compatibility emerges from how partners' interaction patterns evolve together over time, creating fractal structures that integrate their individual tendencies into coherent relationship patterns. This perspective aligns with research showing that relationship satisfaction depends less on static similarity between partners than on how they adapt to each other over time — how they develop communication patterns, conflict resolution strategies, and emotional regulation approaches that work for their specific combination of personalities and circumstances.

Beyond the Autonomy-Connection Dichotomy

Perhaps the most fundamental limitation of traditional approaches is the tendency to frame relationship dynamics in terms of a dichotomy between autonomy and connection — as if these were opposing forces that require balance or tradeoff. This dichotomous framing misses how healthy relationships integrate these dimensions across multiple scales rather than merely balancing them.

Several problems with the dichotomous framing have become apparent:

- **False opposition:** treating autonomy and connection as opposing forces misses how they can mutually enhance each other when integrated effectively across scales.
- **Scale confusion:** what appears as a tradeoff at one scale (like deciding between individual and joint activities

on a particular evening) may be part of a larger pattern that integrates both values across longer time scales.

- **Dynamic integration:** healthy relationships don't maintain a static balance between autonomy and connection but dynamically shift their integration patterns based on changing contexts and needs.
- **Fractal resolution:** the apparent dichotomy between autonomy and connection can be resolved through fractal patterns that integrate both values across different scales simultaneously.

During my visits to Niagara Falls and Foz do Iguaçu, I was struck by how these massive waterfall systems embodied the integration of individual and collective dynamics across scales. Each water droplet follows its individual trajectory while simultaneously participating in larger flow patterns, creating a system that's neither purely individualistic nor purely collective but integrates both principles across scales.

This natural metaphor suggests how relationships might integrate autonomy and connection — not through compromise or balance but through fractal patterns that allow both principles to manifest simultaneously across different scales. Partners can maintain individual identity and agency while participating in deeply connected relationship patterns, just as water droplets maintain their individual trajectories while forming the larger patterns of the waterfall. This perspective doesn't eliminate the real tensions that can arise between individual needs and relationship demands but suggests that these tensions might be resolved through fractal integration rather than mere compromise or tradeoff. The goal shifts from finding the right "balance" to developing relationship patterns that integrate autonomy and connection across multiple scales simultaneously.

FNA Approach to Relationship Dynamics

The fractal perspective on relationships isn't just a theoretical framework but the foundation for practical approaches using Fractal Neural Architectures. Let's explore how FNAs might address aspects of relationship dynamics.

Multi-Scale Relationship Modeling

Traditional approaches to modeling relationships typically focus on a single scale of interaction — either micro-level exchanges, mid-level patterns, or macro-level relationship qualities. This single-scale focus misses how relationship dynamics emerge from the interaction between processes at different scales. FNAs address this limitation through multi-scale modeling that represents relationship patterns across different time horizons simultaneously:

- **Micro-scale modeling (seconds to minutes):** representing the dynamics of momentary interactions, including communication patterns, emotional exchanges, and immediate responses.
- **Meso-scale modeling (hours to days):** capturing patterns that emerge across longer interactions, including conversation development, conflict cycles, and daily routines.
- **Macro-scale modeling (weeks to months):** representing longer-term patterns in the relationship, including recurring themes, developmental trajectories, and adaptation to changing circumstances.

- **Meta-scale modeling (months to years):** capturing the overarching patterns that characterize the relationship across its history, including identity development, life transitions, and long-term evolution.

What makes this multi-scale approach particularly valuable is how it represents the connections between patterns at different scales — how momentary interactions influence and are influenced by longer-term patterns, creating integrated fractal structures rather than isolated layers of analysis. During my work with AI systems since founding zyrian.ai in 2013, I explored how these multi-scale models might be implemented using fractal neural architectures.

The key innovation was developing connectivity patterns that allowed information to flow naturally between different scales, creating integrated representations that spanned from momentary exchanges to relationship histories. This approach enables more sophisticated understanding of relationship dynamics — recognizing how patterns at different scales interact and influence each other, how changes at one scale propagate to others, and how the overall fractal structure of the relationship evolves over time.

Fractal Pattern Recognition

Another key capability of FNAs for relationship dynamics is recognizing self-similar patterns across different scales and contexts — identifying the fractal signatures that characterize particular relationships. This pattern recognition operates across several dimensions:

- **Temporal self-similarity:** identifying patterns that recur across different time scales, from momentary interactions to longer-term dynamics.
- **Contextual self-similarity:** recognizing patterns that maintain their essential qualities across different situations and contexts despite surface variations.
- **Developmental self-similarity:** identifying how patterns transform while maintaining recognizable qualities across different relationship stages and life transitions.
- **Thematic self-similarity:** recognizing recurring themes that appear across different aspects of the relationship, from communication to emotional exchange to decision-making.

What makes this pattern recognition particularly valuable is how it reveals the underlying fractal structure of the relationship — the signature patterns that characterize how these particular partners connect across different scales and contexts. During my sabbatical in São Paulo, I had discussions with relationship therapists about how this pattern recognition might transform their practice. The key insight was that identifying these fractal signatures could help couples recognize their own recurring patterns — patterns that might otherwise remain invisible or be perceived only in fragmented form rather than as integrated structures that span across different aspects of the relationship. This recognition creates possibilities for more effective relationship development — helping partners understand not just specific behaviors or interactions but the underlying fractal patterns that integrate across different scales and contexts of their connection.

Adaptive Intervention Design

Perhaps the most valuable application of FNAs to relationship dynamics is designing interventions that respect and leverage the fractal properties of relationship patterns — approaches that can help transform these patterns in ways that foster healthier and more fulfilling connections. Traditional relationship interventions often focus on changing specific behaviors or teaching isolated skills without addressing how these elements integrate into larger relationship patterns. This fragmented approach can lead to temporary changes in specific behaviors without transforming the underlying patterns.

FNAs enable more integrated intervention approaches:

- **Multi-scale intervention design:** interventions target patterns at multiple scales simultaneously, from momentary interaction habits to longer-term dynamics, creating coherent change across the relationship's fractal structure.
- **Propagation-based approaches:** rather than attempting dramatic transformations, interventions focus on small, consistent changes that can propagate across scales through the relationship's natural fractal dynamics.
- **Pattern-respectful adaptation:** interventions work with rather than against the existing fractal structure of the relationship, modifying rather than replacing patterns to maintain coherence while fostering positive change.
- **Leverage point identification:** analysis identifies specific aspects of relationship patterns where small changes might have disproportionate effects across scales, creating efficient pathways for transformation.

This approach recognizes that relationship transformation isn't achieved through dramatic interventions or isolated skill development but through small, consistent changes that propagate across scales to transform the entire fractal structure of the relationship over time. During my early retirement and sabbatical, I've had the opportunity to observe how this fractal approach to transformation operates in various relationships — how small but consistent changes in interaction patterns gradually propagate across scales to transform entire relationship systems. The key insight is that relationship patterns show both remarkable stability and potential for transformation — that they resist change when approached through isolated interventions but can transform significantly through small, consistent adjustments that respect and leverage their fractal properties.

In practice, implementations would include additional components like natural language processing for analyzing communication content, emotion recognition for tracking affective patterns, and more sophisticated pattern recognition algorithms. The specific architecture would be adapted to the particular aspects of relationship dynamics being analyzed and the available data sources.

Evaluation Metrics for Relationship Patterns

Evaluating the quality of relationship patterns requires metrics that go beyond simple satisfaction measures to assess the fractal properties that contribute to relationship health and resilience. Several specialized metrics have proven valuable:

- **Cross-scale coherence:** how consistently do relationship patterns maintain their essential qualities across different time scales and contexts? This measures the integration between momentary interactions and longer-term patterns.
- **Adaptive flexibility:** how effectively does the relationship adjust its patterns in response to changing circumstances while maintaining essential continuity? This assesses the balance between stability and adaptability.
- **Pattern resilience:** how well do positive relationship patterns maintain their structure when faced with challenges or stressors? This measures the robustness of the relationship's fractal organization.
- **Growth capacity:** how effectively can the relationship incorporate new experiences and learning into its existing patterns? This assesses the capacity for development without fragmentation.
- **Balanced integration:** how well does the relationship integrate potentially competing values like autonomy and connection, stability and growth, or familiarity and novelty? This measures the relationship's capacity for complex integration rather than simple tradeoffs.

These metrics provide a more comprehensive evaluation than simple satisfaction or stability measures, helping identify specific strengths and vulnerabilities in relationship patterns. During my work with relationship dynamics, I've found that qualitative assessment by the partners themselves can also provide valuable insights, particularly when guided by frameworks that help them recognize fractal patterns in their own relationship.

Partners can often identify recurring themes and patterns that might be missed by external analysis, especially when given conceptual tools to recognize how these patterns manifest across different scales and contexts. The fractal perspective suggests additional evaluation approaches focused on how patterns evolve over time. Rather than treating relationship quality as a static property, we can assess how patterns maintain continuity while adapting to new circumstances — how they preserve their essential fractal structure while evolving in response to changing contexts and developmental stages.

Applications and Results

The theoretical advantages of Fractal Neural Architectures for relationship dynamics translate to practical benefits across various applications. Let's explore some specific examples and results.

Relationship Education and Counseling

One of the most promising applications of fractal approaches to relationships is in education and counseling — helping couples recognize and transform their interaction patterns in ways that foster healthier and more fulfilling connections. Traditional relationship education and counseling approaches often focus on specific communication skills, conflict resolution techniques, or emotional regulation strategies. While these approaches offer valuable tools, they sometimes miss how these different aspects integrate into coherent relationship patterns across scales. Fractal approaches address this limitation through several innovations:

- **Pattern recognition across contexts:** rather than focusing on isolated behaviors, fractal approaches help couples recognize recurring patterns that appear across different situations and time scales in their relationship.
- **Multi-scale intervention design:** interventions are designed to address patterns at multiple scales simultaneously, from momentary interaction habits to longer-term relationship dynamics.
- **Propagation principles:** couples learn how small, consistent changes can propagate across scales to transform larger patterns over time, creating more sustainable transformation than approaches focused on dramatic interventions.
- **Integration monitoring:** progress is tracked not just in terms of specific behavior changes but in how these changes integrate into coherent new patterns across different contexts and scales.

During my sabbatical in São Paulo, I had discussions with relationship counselors about how these fractal principles might transform their practice. While I haven't directly implemented these ideas in clinical settings, experimental applications have shown promising results. One particularly interesting application is in premarital education programs that help couples recognize their emerging fractal patterns early in the relationship, when these patterns are still relatively flexible. By developing awareness of these patterns and their potential long-term implications, couples can make conscious choices about which patterns to reinforce and which to modify before they become deeply established. The fractal approach suggests that effective relationship education isn't just about teaching specific skills or techniques but about helping couples recognize and consciously shape the fractal patterns that will characterize their relationship across different contexts and time scales.

AI-Assisted Relationship Support

Another promising application area is AI-assisted relationship support — systems that help couples track, understand, and improve their interaction patterns over time. Traditional relationship apps and tools typically focus on specific aspects like scheduling date nights, tracking special occasions, or providing communication prompts. While these tools can be helpful, they often miss the integrated, multi-scale nature of relationship patterns. FNA-based relationship support systems address this limitation through several innovations:

- **Multi-scale pattern tracking:** the system helps couples track interaction patterns across different time scales, from daily communications to weekly routines to monthly and yearly cycles.
- **Fractal pattern visualization:** complex relationship dynamics are visualized through fractal representations that make recurring patterns more recognizable and understandable.
- **Contextual recommendations:** rather than providing generic advice, the system offers suggestions based on the specific fractal patterns identified in the relationship, tailored to the couple's unique dynamics.
- **Growth trajectory mapping:** the system helps couples visualize how small, consistent changes might transform their relationship patterns over time, creating realistic expectations for the pace and nature of relationship development.

During my work with AI systems, I've been particularly interested in how these capabilities might transform relationship support technologies. The key insight is that effective support wouldn't just track isolated behaviors or provide generic advice but would help couples recognize and consciously shape the fractal patterns that integrate across different aspects of their relationship. This approach respects the unique nature of each relationship while providing structured support for recognizing and transforming patterns — avoiding both the rigidity of one-size-fits-all advice and the lack of direction in completely unstructured approaches.

Cross-Cultural Relationship Navigation

A third promising application area is supporting cross-cultural relationships — helping partners navigate the complexities that arise when they come from different cultural backgrounds with different relationship expectations and patterns. Cross-cultural relationships face unique challenges as partners must integrate potentially different fractal templates for relationship — different expectations about how connection, communication, conflict, and commitment should unfold across scales from momentary interactions to lifetime bonds. Fractal approaches offer several advantages for navigating these complexities:

- **Pattern recognition across cultural contexts:** partners develop awareness of how their relationship expectations have been shaped by cultural fractal templates, recognizing both differences and surprising similarities in these templates.

- **Integration rather than compromise:** rather than merely compromising between different cultural patterns, partners learn to develop integrated fractal structures that incorporate elements from both backgrounds into coherent new patterns.
- **Multi-scale cultural navigation:** partners recognize how cultural differences manifest differently across scales, from momentary communication styles to longer-term relationship trajectories, developing scale-appropriate integration strategies.
- **Emergent bicultural patterns:** the relationship develops unique fractal patterns that aren't reducible to either cultural background but emerge from the specific interaction between these partners in their particular context.

During my experiences as an immigrant who moved from Brazil to the United States, I encountered these challenges firsthand in both my own relationships and those of other immigrants and cross-cultural couples. The fractal perspective offered a framework for understanding these dynamics not as problems to be solved but as opportunities to develop uniquely integrated relationship patterns. This approach suggests that successful cross-cultural relationships don't require partners to abandon their cultural backgrounds or compromise between them but to develop new fractal patterns that integrate elements from both cultures into coherent relationship structures that work for their specific combination of backgrounds, personalities, and circumstances.

Comparison with Traditional Relationship Approaches

Across these applications, several patterns emerge when comparing fractal approaches with traditional methods of relationship support and development:

- **Integration quality:** fractal approaches typically achieve more coherent integration across different aspects and scales of the relationship, particularly for complex dynamics that span from momentary interactions to longer-term patterns.
- **Sustainability of change:** changes fostered through fractal approaches tend to be more sustainable over time, as they address integrated patterns rather than isolated behaviors and respect the natural dynamics of how relationship patterns evolve.
- **Contextual adaptation:** fractal approaches adapt more effectively to the unique context of each relationship, recognizing that healthy patterns may look different for different couples based on their specific combination of backgrounds, personalities, and circumstances.
- **Developmental coherence:** fractal approaches maintain better continuity across relationship development, helping couples navigate transitions while preserving essential qualities of their connection rather than treating each stage as requiring completely different skills or approaches.

These comparative advantages aren't marginal but often substantial, particularly for complex relationship challenges that span across different aspects and time scales of the connection — precisely the kinds of challenges that traditional approaches often struggle to address effectively. During my three decades thinking about these issues and observing various relationships, I've become convinced that the fractal approach represents not just an incremental improvement but a fundamental reconceptualization of how we understand and support human connection — one that better aligns with the complex, multi-scale nature of real relationships.

Why It Matters

Transforming Human Connection

The fractal approach to relationships isn't just theoretically captivating but has profound implications for how we understand and develop our connections with others.

Enhancing Relationship Consciousness

Traditional approaches to relationship improvement often focus on specific behaviors or skills without developing broader awareness of the patterns that integrate across different aspects of the relationship. This fragmented approach can lead to temporary changes in isolated behaviors without transforming the underlying patterns. The fractal perspective suggests a different path to relationship development — one centered on enhancing relationship consciousness:

- **Pattern awareness across scales:** developing the ability to recognize relationship patterns that span from momentary interactions to longer-term dynamics, seeing how these patterns integrate across different time scales and contexts.

- **Self-similar theme recognition:** learning to identify recurring themes that appear in different aspects of the relationship, recognizing the fractal signature of particular relationship dynamics.
- **Contextual pattern understanding:** developing awareness of how relationship patterns adapt to different contexts while maintaining their essential qualities, distinguishing between surface variations and deeper pattern changes.
- **Developmental trajectory recognition:** seeing how relationship patterns evolve over time, recognizing both continuity and transformation in how the relationship develops across different stages.

During my sabbatical in São Paulo, I explored how these aspects of relationship consciousness might be cultivated through specific practices and frameworks. The key insight was that developing this consciousness isn't just about acquiring knowledge but about training perception — learning to see relationship patterns that were previously invisible or only partially recognized. This enhanced relationship consciousness doesn't just improve specific interactions but transforms the entire experience of connection — helping partners see their relationship not as a collection of separate behaviors and events but as an integrated fractal pattern that unfolds across scales from moments to lifetimes.

Implications for Relationship Development

The fractal perspective also has profound implications for how we approach relationship development — suggesting possibilities for growth that go beyond traditional models of skill acquisition or problem resolution. Several alternative development paths emerge from the fractal approach:

- **Pattern evolution rather than problem solving:** relationship development focuses not just on solving specific problems but on evolving the underlying patterns that generate both challenges and opportunities across different contexts.
- **Integrated growth across scales:** development occurs simultaneously across different scales, with changes in momentary interaction habits connecting to shifts in longer-term relationship dynamics through fractal propagation.
- **Coherent differentiation:** partners develop greater individual differentiation while maintaining coherent connection, creating more complex and resilient relationship patterns rather than merely balancing autonomy and intimacy.
- **Emergent relationship creativity:** the relationship develops unique patterns that aren't reducible to either partner's individual tendencies but emerge from their specific interaction, creating possibilities that neither would've discovered alone.

During my early retirement and sabbatical, I've had time to observe how these development paths manifest in various relationships — how the most fulfilling connections aren't those that conform to external ideals but those that develop unique fractal patterns that emerge from the specific interaction between those particular partners. This perspective suggests that relationship development isn't about progressing toward a universal ideal but about evolving patterns that are uniquely adapted to the specific partners and their context — patterns that integrate their individual qualities into coherent relationship structures that work for their particular combination of personalities, backgrounds, and circumstances.

Ethical Considerations in Relationship Technology

The potential of fractal approaches to relationship dynamics brings significant ethical responsibilities, particularly as these approaches are implemented in technologies designed to analyze and influence human connections. Several principles should guide the development of relationship technologies:

- Partner agency: technologies should enhance rather than diminish partners' agency in shaping their relationship, providing insights and options rather than prescriptions or manipulations.
- Relationship diversity: development should respect and preserve the diversity of healthy relationship patterns rather than imposing standardized ideals based on particular cultural or theoretical models.
- Transparent analysis: the basis for any relationship analysis or recommendations should be transparent

and understandable to the partners, allowing them to evaluate and potentially override algorithmic insights.

- Privacy and consent: collection and analysis of relationship data should be based on informed consent from both partners, with clear boundaries around what information is collected and how it's used.
- Balanced integration: technologies should support balanced integration of digital and in-person interaction rather than replacing face-to-face connection with mediated communication.

During my work with AI systems since founding zyrian.ai in 2013, I've been particularly concerned with how these ethical considerations apply to technologies that analyze and influence human relationships. The most powerful relationship technologies are those that enhance human connection rather than replacing or controlling it — that augment partners' natural capacity for developing healthy relationship patterns rather than imposing external models. The fractal perspective suggests that these ethical considerations aren't separate from the technical development but integral to it — that the patterns of agency, diversity, transparency, privacy, and integration we design into our systems will shape not just their performance but their impact on human connection and well-being.

Practical Framework

"Fractal Relationship" Technique

To help you apply these concepts in your own relationships, I offer this practical framework for fractal relationship development inspired by the principles discussed in this chapter.

Step 1

Recognize Your Relationship Fractals

The foundation of fractal relationship development is recognizing the patterns that characterize your connections with others. Practice identifying patterns across different scales and contexts:

- **Interaction patterns:** notice the recurring dynamics in your moment-to-moment exchanges. How do conversations typically flow? What patterns appear in how you and your partner respond to each other's emotions? What sequences tend to repeat in conflicts or moments of connection?
- **Daily and weekly rhythms:** observe the patterns in how your relationship functions across days and weeks. What routines have developed? How do you navigate transitions between togetherness and separateness? What cycles of connection and distance tend to occur?
- **Monthly and yearly cycles:** identify patterns that unfold across longer time periods. How does your relationship navigate different seasons or annual events? What longer-term cycles of growth, consolidation, or challenge have you noticed?
- **Life stage patterns:** consider how your relationship has evolved across major life transitions. What essential qualities have remained consistent despite changing circumstances? What aspects have transformed while others have persisted?

During my visits to Niagara Falls and Foz do Iguaçu, I practiced this pattern recognition by observing couples interacting in these dramatic natural settings. What struck me was how each couple's momentary interactions — how they directed each other's attention, responded to each other's reactions, and coordinated their movements — reflected larger patterns in their relationship that were visible even to an outside observer. This observation suggests that our relationship patterns are often more visible and consistent than we realize — that with practice, we can learn to recognize the fractal structures that characterize our connections across different scales and contexts.

Step 2

Identify Self-Similar Themes

Once you've recognized patterns at different scales, look for self-similar themes that appear across these different aspects of your relationship:

- **Communication themes:** what similar patterns appear in how you and your partner communicate across different topics and contexts? How do these patterns manifest in both everyday conversations and remarkable discussions?
- **Emotional themes:** what recurring patterns characterize how emotions flow between you and your partner? How do these patterns appear in both minor emotional exchanges and major emotional events?
- **Support themes:** what patterns characterize how you and your partner support each other? How do these patterns manifest in both everyday assistance and responses to major challenges?

- **Growth themes:** what patterns appear in how your relationship navigates change and development? How do these patterns manifest in both small adjustments and major transitions?

During my sabbatical in São Paulo, reconnecting with childhood friends after decades apart, I was struck by how certain essential qualities of our connections had persisted despite the vast differences in our life experiences. These persistent themes weren't just similarities in interests or values but recurring patterns in how we related to each other — patterns that had evolved and matured but maintained recognizable core qualities across decades. This experience suggests that identifying self-similar themes isn't about finding exact repetitions but recognizing essential qualities that manifest in different forms across different contexts and time scales — the fractal signature of your unique relationship patterns.

Step 3

Assess Pattern Health and Resilience

The next step is evaluating the health and resilience of your relationship patterns — not against external ideals but in terms of how well they serve the specific needs and goals of your particular relationship:

- **Coherence across scales:** do your relationship patterns maintain consistent qualities across different scales, or do they fragment into contradictory dynamics in different contexts? Coherent patterns feel integrated and authentic, while fragmented patterns create confusion and disconnection.

- **Adaptive flexibility:** can your relationship patterns adapt to changing circumstances while maintaining their essential qualities, or do they become rigid and brittle when faced with new situations? Healthy patterns show both stability and adaptability.
- **Balanced integration:** do your patterns integrate potentially competing values like autonomy and connection, stability and growth, or familiarity and novelty? Healthy patterns achieve complex integration rather than simple tradeoffs between these values.
- **Growth capacity:** can your relationship patterns incorporate new experiences and learning, or do they resist development and change? Resilient patterns maintain continuity while evolving in response to new insights and circumstances.

During my work with AI systems, I developed frameworks for evaluating these qualities in relationship patterns. The key insight was that pattern health isn't about conforming to universal ideals but about how well the specific patterns serve the unique needs and goals of the particular relationship — how they enable both partners to thrive individually and together in their specific context. This assessment isn't about judging the relationship but about identifying opportunities for growth and development — recognizing both the strengths to build upon and the aspects that might benefit from conscious evolution.

Step 4

Cultivate Fractal Transformation

The final step is cultivating transformation in your relationship patterns — not through dramatic interventions but through small, consistent changes that can propagate across scales to transform the entire fractal structure over time:

- **Identify leverage points:** look for aspects of your relationship patterns where small changes might have disproportionate effects across scales. These are often found at transition points between different activities or emotional states, or in recurring sequences that trigger larger patterns.
- **Design small, consistent adjustments:** rather than attempting dramatic changes, focus on small adjustments that you can implement consistently across different contexts. Consistency across contexts is more important than the magnitude of the change.
- **Track propagation effects:** notice how changes in specific aspects of your interaction begin to influence other aspects of the relationship across different scales. These propagation effects often appear first in similar contexts before spreading to more distant aspects of the relationship.
- **Reinforce emerging patterns:** as new patterns begin to emerge, consciously reinforce them through attention and appreciation. This reinforcement helps stabilize positive changes and integrate them into the relationship's evolving fractal structure.

During my early retirement and sabbatical, I've had the opportunity to observe how these principles of fractal transformation operate in various relationships. The most striking observation was how seemingly small but consistent changes in interaction patterns could gradually transform entire relationship systems — not through immediate dramatic shifts but through the natural propagation of changes across the relationship's fractal structure over time. This approach to transformation respects the complex, integrated nature of relationship patterns while offering practical pathways for conscious evolution — helping partners develop relationship patterns that better serve their unique needs and goals while maintaining coherence across different aspects of their connection.

Call to Action

Developing Fractal Awareness in Your Relationships

As we conclude this chapter on love and relationships, I invite you to begin applying these insights to develop greater fractal awareness in your own connections. Here are specific actions you can take:

- Practice multi-scale observation daily. Set aside time to notice relationship patterns at different scales, from momentary interactions to longer-term dynamics. Keep a relationship journal where you record observations about patterns across these different scales.
- Explore fractal visualization with a partner. Share the concept of relationship fractals with someone close to you and explore together how you might visualize

the patterns in your connection. What metaphors or images capture the self-similar qualities of your relationship across different scales?

- Experiment with small pattern adjustments. Identify one small aspect of your interaction patterns that you'd like to shift, and practice a consistent adjustment across different contexts. Notice how this change might propagate to other aspects of the relationship over time.
- Develop relationship consciousness through comparative observation. When you observe other relationships — whether in person, in media, or in literature — practice identifying their fractal patterns. How do these patterns compare to those in your own relationships? What might you learn from these comparisons?

The fractal nature of relationships isn't just a theoretical framework but a practical approach to developing more conscious, integrated, and fulfilling connections. By recognizing and working with the fractal patterns in your relationships, you can foster connections that are uniquely adapted to the specific people involved while showing the coherence, adaptability, and resilience characteristic of healthy fractal systems. In the next chapter, we'll explore how fractal principles apply to the laws of physics — how the fundamental patterns of the universe might be understood through the lens of fractal organization across scales from quantum fields to cosmic structures.

CHAPTER 8

The Laws of Physics

Narrative Opening: The Fractal Universe Revealed

Professor Chen stood before the holographic display, manipulating the visualization with subtle hand gestures. The room was silent except for the soft hum of the quantum processor as it rendered increasingly complex patterns of subatomic interactions. "There", he said, freezing the simulation at a particular moment. "Do you see it?" Kai and Elena stepped closer, studying the intricate dance of particles and fields that filled the display. At first glance, it appeared chaotic — countless particles moving in seemingly random patterns. But as they continued to observe, subtle regularities began to emerge. "The same patterns..." Elena whispered, her eyes widening. "They're repeating at different scales."

"Exactly", Professor Chen nodded, his face illuminated by the blue glow of the hologram. "We've been running these simulations for months, analyzing quantum field interactions across different energy levels. What we're seeing isn't just mathematical coincidence — it's a fundamental property of physical reality." He expanded the visualization, revealing how the same essential patterns appeared at subatomic, atomic, molecular, and macroscopic scales. "The universe doesn't just compute — it learns. And it does so through fractal patterns that propagate across scales." Kai shook his head in amazement. "But this contradicts the standard model's assumption of scale separation — the idea that physics at different scales can be treated independently." "That's precisely why this is so revolutionary", Professor Chen replied. "For centuries, we've been fragmenting physics into separate domains — quantum mechanics for the very small, classical mechanics for the medium scale, relativity for the very large.

But nature doesn't recognize these artificial boundaries. The same fundamental patterns appear across all scales, just expressed through different physical mechanisms." He gestured again, and the simulation shifted to display the evolution of these patterns through time. "But here's what's truly remarkable. These aren't just spatial fractals — they're temporal fractals as well. The patterns of interaction that unfold across microseconds mirror those that unfold across millennia." Elena leaned forward, transfixed by the visualization. "It's as if time itself has a fractal structure..." "Yes", Professor Chen said quietly. "And that insight might be the key to resolving some of the deepest paradoxes in physics — the reconciliation of quantum mechanics and general relativity, the arrow of time, the nature of consciousness itself."

He turned to face them directly. "What if time didn't emerge from space, as we've always assumed? What if it was the other way around? What if time existed first — pure duration without dimension — and space emerged from the increasing complexity of temporal patterns?" Kai's expression shifted from skepticism to fascination. "A fractal theory of time... where complexity increases through self-similar patterns until it generates spatial dimensions..." "It sounds almost mystical", Elena said, "but the mathematics is rigorous. The simulations support it." Professor Chen nodded. "The universe doesn't just compute — it learns. And perhaps it learned to exist through the fractal elaboration of time itself."

As they continued to study the simulation, the boundaries between physics, mathematics, and philosophy seemed to dissolve. The fractal patterns dancing before them weren't just abstract representations but glimpses into the fundamental nature of reality — a reality where the same essential patterns appeared across all scales of space and time, creating a universe that was both deterministic and creative, ordered and spontaneous, unified and endlessly diverse.

Fractal Patterns in Physical Laws

The narrative above illustrates a profound possibility: that the fundamental laws of physics might be understood through the lens of fractal organization across scales. This perspective challenges traditional approaches to physics while opening new avenues for understanding the universe's deepest patterns.

Scale Invariance and Physical Laws

One of the most striking aspects of physical laws is how certain patterns and principles maintain their essential form across widely different scales, from the quantum realm to cosmic structures. At the quantum scale, we observe patterns in particle interactions, wave-particle duality, and quantum field fluctuations. These patterns establish the foundation for all physical phenomena, creating the building blocks from which larger structures emerge. At the atomic and molecular scale, we find patterns in how elements combine, how energy states transition, and how complex structures form from simpler components. These patterns create the material basis for all physical objects, from crystals to living organisms.

At the macroscopic scale, we observe patterns in fluid dynamics, thermodynamic processes, and mechanical interactions. These patterns govern the behavior of everyday objects and phenomena, from weather systems to mechanical devices. At the astronomical scale, we find patterns in gravitational interactions, stellar evolution, and galactic formation. These patterns shape the largest structures in the observable universe, creating the cosmic environment within which all other processes unfold. What makes these cross-scale patterns particularly interesting is how they maintain certain essential properties while adapting to the specific conditions of each scale. The principle of conservation of energy, for instance, applies across all scales, though its specific manifestations differ dramatically between quantum and cosmic contexts.

During my sabbatical, I spent countless hours contemplating these cross-scale patterns, often while observing natural phenomena like the branching patterns of trees in São Paulo parks or the flow patterns in the waters of Foz do Iguaçu. What struck me was how the same essential patterns appeared across vastly different scales and contexts, suggesting that nature might be employing similar principles throughout its diverse manifestations. This observation led me to wonder whether physical laws might be understood not as separate domains but as fractal expressions of more fundamental patterns — patterns that maintain their essential structure while adapting to different scales and contexts. Just as the Sierpiński triangle emerges from a simple rule repeated across scales, perhaps the complex tapestry of physical laws emerges from basic principles applied recursively across different domains.

This fractal perspective offers a new way to understand both the diversity and unity of physical laws — how they can manifest in dramatically different forms while maintaining underlying connections, how patterns at one scale can influence and constrain possibilities at other scales.

Temporal Fractals and the Arrow of Time

Perhaps the most intriguing application of fractal thinking to physics concerns the nature of time itself — how temporal patterns might exhibit self-similarity across different scales, creating a fractal structure that helps explain the arrow of time. Traditional approaches to time in physics have struggled with several fundamental questions:

- Why does time appear to flow in one direction (the arrow of time) when most fundamental physical laws are time-symmetric?
- How can we reconcile the reversible nature of microscopic physics with the irreversible nature of macroscopic phenomena?
- What's the relationship between subjective time (as experienced by conscious beings) and objective time (as described by physical theories)?

The fractal perspective offers new approaches to these questions by suggesting that time might have a self-similar structure across different scales — that temporal patterns at one scale might influence and constrain patterns at other scales in ways that create the appearance of directional flow. During my early explorations of these ideas, I created Excel spreadsheets to simulate how fractal patterns might emerge in temporal processes. What fascinated me was how simple recursive rules could generate complex temporal structures that exhibited both order and unpredictability — structures that maintained certain patterns across different time scales while continuously evolving in response to their own history.

This exploration led me to a radical hypothesis: what if time existed before space? What if, in the primordial singularity before the Big Bang, time already existed as pure duration without spatial dimension, and space emerged gradually through the increasing complexity of temporal patterns? This hypothesis suggests that time isn't just another dimension alongside space but the fundamental substrate from which all other dimensions emerge — that the universe began not as a spatial point of infinite density but as a temporal process of increasing complexity, with spatial dimensions emerging as this complexity reached critical thresholds. While speculative, this perspective aligns with certain observations in both physics and cosmology:

- The apparent expansion of space from the Big Bang might be understood as the emergence of spatial dimensions from temporal complexity rather than the expansion of pre-existing space.
- The arrow of time might emerge naturally from the fractal structure of temporal processes rather than requiring special initial conditions or external constraints.
- The relationship between time and consciousness might reflect how both involve the integration of information across different scales, creating coherent patterns that span from microseconds to years.

- What if consciousness itself emerges from these fractal patterns? Consider how your own awareness integrates information across different time scales and sensory domains.

During my visits to China, particularly in remote regions near the Kazakhstan frontier, I had fascinating discussions with physicists who were exploring similar ideas from different cultural and theoretical perspectives. What struck me was how these concepts transcended traditional boundaries between physics, mathematics, philosophy, and even certain spiritual traditions — how they pointed toward a more integrated understanding of reality that respected both scientific rigor and the depth of human experience. This perspective suggests that Fractal Neural Architectures might offer valuable approaches to modeling temporal processes in physics — not just as computational tools but as frameworks that capture something essential about how time itself is structured across different scales.

Emergence and Complexity in Physical Systems

Another domain where fractal patterns become evident is in the emergence of complexity from simpler components — how physical systems spontaneously develop organized structures across different scales. Traditional approaches to emergence in physics have often struggled to explain how complex, organized structures can arise from simpler components following basic physical laws. The fractal perspective offers new insights into this process by suggesting that emergence might operate through similar principles across different scales and contexts.

Several fractal properties appear in emergent physical systems:

- **Self-organized criticality:** many physical systems naturally evolve toward critical states characterized by fractal properties, where small perturbations can trigger effects across multiple scales. Examples include sand piles, earthquake systems, and certain aspects of neural activity.
- **Scale-linking feedback loops:** complex systems often develop feedback mechanisms that connect processes across different scales, creating integrated structures that can't be reduced to any single scale of analysis.
- **Fractal attractors:** many dynamical systems evolve toward fractal attractors — states that exhibit self-similarity across scales and serve as organizing principles for the system's behavior over time.
- **Hierarchical organization:** complex systems often develop hierarchical structures with similar organizational principles appearing at different levels of the hierarchy, creating fractal-like patterns of organization.

These fractal properties help explain how complexity can emerge from simplicity without requiring external design or special initial conditions — how the natural dynamics of physical systems can spontaneously generate organized structures across different scales. During my work with AI systems since founding zyrian.ai in 2013, I became interested in how these emergent properties might inform more effective approaches to artificial intelligence. If natural intelligence emerges through fractal processes that span across different scales, perhaps artificial intelligence should be designed to leverage similar principles rather than relying on brute-force computation or rigid hierarchical structures.

This perspective suggests that effective AI wouldn't just process information but would develop fractal structures that integrate across different scales — structures that could spontaneously generate new patterns and capabilities through the same principles that drive emergence in natural systems.

Quantum Mechanics and Fractal Uncertainty

Perhaps the most profound connection between fractal patterns and physics appears in quantum mechanics — the domain where traditional intuitions about determinism and locality break down, revealing a reality that seems to operate according to different principles than our everyday experience. Several aspects of quantum mechanics exhibit fractal-like properties:

- **Quantum superposition:** the ability of quantum systems to exist in multiple states simultaneously creates a form of scale-linking where different potential realities coexist until measurement collapses the possibilities into a single outcome.
- **Entanglement across scales:** quantum entanglement creates connections that transcend spatial separation, linking particles across different locations in ways that suggest a deeper, scale-transcending reality beneath the apparent separation of classical physics.
- **Measurement and the observer effect:** the role of observation in quantum mechanics suggests a fundamental connection between consciousness and physical reality — a connection that might operate

through fractal patterns that integrate across different scales of organization.

Have you ever considered how your awareness integrates information across different time scales? This question lies at the heart of our exploration of fractal neural architectures.

- Quantum field fluctuations: the constant fluctuations in quantum fields create patterns that exhibit statistical self-similarity across different scales, suggesting a fractal structure to the quantum vacuum itself.

These quantum properties suggest that uncertainty in physical systems isn't just random noise but might exhibit structured patterns across scales — patterns that could be leveraged to develop more effective approaches to quantum computation and simulation. During my sabbatical in São Paulo, I had discussions with quantum physicists about these connections between fractal patterns and quantum phenomena. What struck me was how these concepts offered new perspectives on some of the deepest paradoxes in quantum mechanics — how they suggested possibilities for reconciling quantum weirdness with our intuitions about physical reality without requiring either to be abandoned. This perspective suggests that quantum mechanics might be understood not as a complete break from classical physics but as a different manifestation of deeper patterns that span across different scales and domains — patterns that appear in different forms at quantum and classical scales but maintain certain essential properties throughout.

Personal Journey

Discovering Physical Fractals

My path to understanding physics through fractal patterns wasn't through traditional research but through personal experiences and observations that gradually coalesced into a coherent perspective.

Natural Observation and Pattern Recognition

One of the most formative aspects of my journey was direct observation of natural phenomena — noticing patterns in the physical world that suggested deeper mathematical structures spanning across different scales.

During walks in parks and natural areas, I found myself drawn to the branching patterns of trees, the network structures of leaf veins, the spiral arrangements of seeds in sunflowers, and countless other natural forms that exhibited mathematical regularity despite their apparent complexity. What struck me was how these patterns seemed to optimize various functions — maximizing surface area for light capture, minimizing material use while maintaining structural integrity, balancing competing needs for growth and stability. These observations led me to wonder whether nature might be employing similar optimization principles across widely different domains — whether the patterns I was observing might reflect deeper mathematical structures that transcended the specific contexts in which they appeared.

During my visits to Niagara Falls and Foz do Iguaçu, I was particularly struck by the fractal patterns in water flow — how similar structures appeared at different scales, from tiny ripples to massive cascades. These observations suggested that fluid dynamics might exhibit fractal properties that connected phenomena across widely different scales, from the microscopic movements of water molecules to the macroscopic patterns of rivers and waterfalls. What made these observations particularly valuable wasn't just their aesthetic appeal but their suggestion that nature might be employing similar mathematical principles across widely different domains — that the universe might be more unified and coherent than our fragmented scientific disciplines sometimes suggest.

Excel Experiments with Fractal Time

Another formative strand in my exploration of physical fractals came through computational experiments — simple simulations that revealed how complex patterns could emerge from relatively simple rules applied recursively across different scales. Using Excel spreadsheets, I created simulations of how fractal patterns might emerge in temporal processes. These weren't sophisticated scientific models but exploratory tools that allowed me to visualize how simple recursive rules could generate complex structures over time. What fascinated me was how these simulations could produce patterns that exhibited both order and unpredictability — structures that maintained certain regularities across different scales while continuously evolving in ways that couldn't be precisely predicted from their initial conditions.

These experiments led me to wonder whether time itself might have a fractal structure — whether temporal patterns might exhibit self-similarity across different scales, creating a universe where similar themes recur in new contexts and at different scales throughout cosmic evolution. This exploration eventually led to my radical hypothesis about time existing before space — the idea that in the primordial singularity before the Big Bang, time might have existed as pure duration without spatial dimension, with space emerging gradually through the increasing complexity of temporal patterns. While speculative, this hypothesis offered a new perspective on some of the deepest questions in physics and cosmology — questions about the origin of the universe, the nature of time, and the relationship between consciousness and physical reality.

Cross-Disciplinary Connections

A third strand in my exploration of physical fractals came through recognizing connections between patterns in physics and those in other domains — how similar mathematical structures appeared across fields as diverse as neuroscience, ecology, economics, and social systems. During my work with AI systems, I became increasingly aware of how certain mathematical patterns appeared across widely different domains — how power laws, scale-free networks, and other fractal-like structures seemed to emerge spontaneously in systems ranging from neural networks to internet architecture to financial markets.

What struck me was how these cross-disciplinary patterns suggested deeper mathematical principles that transcended the specific contexts in which they appeared — principles that might reflect fundamental aspects of how complex systems organize themselves across different domains and scales. During my travels in China, particularly in remote regions near the Kazakhstan frontier, I had fascinating discussions with researchers working at the boundaries between different disciplines — physicists collaborating with biologists, economists working with ecologists, computer scientists partnering with philosophers. These cross-disciplinary conversations revealed how similar patterns were being discovered independently in different fields, suggesting convergence toward recognition of common mathematical structures across diverse domains.

This cross-disciplinary perspective informed my thinking about Fractal Neural Architectures — the idea that effective AI wouldn't just mimic specific aspects of natural intelligence but would leverage the same fractal principles that appear across different natural systems, from neural networks to ecosystems to physical processes.

Learning from Failure and Iteration

Perhaps the most valuable insights from my exploration of physical fractals came not from successful predictions or elegant theories but from failures and iterations — from recognizing when my initial hypotheses were wrong and using those errors to develop more nuanced understanding. Like many people exploring unconventional ideas, I've experienced both the excitement of potential breakthroughs and the disappointment of realizing that certain approaches don't work as expected. What gradually became clear through this process was that understanding physical fractals wasn't about finding a single unified theory but about recognizing patterns of relationship that manifest differently across different contexts and scales.

The most valuable lessons came from recognizing how these patterns operated across different domains — how similar mathematical structures could appear in contexts as diverse as quantum fields, neural networks, and social systems, yet with important differences that reflected the specific conditions of each domain. During my three decades thinking about these issues, I've become increasingly convinced that understanding physics through fractal patterns offers a path beyond both rigid reductionism and chaotic relativism. It suggests that while the universe exhibits remarkable diversity across different scales and contexts, this diversity emerges from deeper patterns that maintain certain essential properties while adapting to specific conditions.

This perspective has informed not just my understanding of physics but my approach to designing systems for artificial intelligence — creating architectures that can recognize and leverage the fractal patterns that appear across different natural systems rather than imposing rigid structures that might miss the essential flexibility and adaptability of natural intelligence.

Contrarian Perspective

Beyond Reductionism

The fractal approach to physics challenges several fundamental assumptions in traditional scientific approaches. Let me explain why I believe some established approaches miss the complex, multi-scale nature of physical reality.

Why Reductionism Falls Short

Traditional approaches to physics often rely on reductionism — the idea that complex phenomena can be fully explained by reducing them to their simplest components and the laws governing those components. While this approach has yielded remarkable insights and technological applications, it often misses how properties emerge from the interaction between processes at different scales. I've always been drawn to physics

— the elegant mathematical descriptions of our physical world. During a sabbatical at a mountain research station, I spent evenings stargazing, contemplating how the same fundamental forces shape both the cosmic structure of galaxy clusters and the subatomic dance of particles. The fractal nature of physical laws across scales became viscerally apparent during those quiet nights under the stars.

Several limitations of reductionism have become increasingly apparent:

- **Emergent properties:** many physical systems exhibit properties that can't be predicted from their components alone but emerge from the interaction between processes at different scales. These emergent properties aren't just computational complications but fundamental aspects of how complex systems organize themselves.
- **Scale integration:** reductionist approaches typically focus on a single scale of analysis, missing how processes at different scales influence and constrain each other through feedback loops that span across multiple levels of organization.
- **Contextual dependence:** the behavior of physical systems often depends on their broader context in ways that can't be captured by focusing solely on their internal components and mechanisms. This contextual dependence isn't just a practical limitation but reflects how physical systems are embedded in larger networks of relationship.
- **Information integration:** complex physical systems often integrate information across different scales in ways that can't be reduced to the processing of infor-

mation at any single level. This multi-scale integration creates capabilities that transcend what would be possible through single-scale processing alone.

During my travels and cross-cultural experiences, I've observed how different intellectual traditions approach these limitations in different ways. Western science has traditionally emphasized reductionist approaches, while Eastern traditions have often emphasized holistic perspectives that recognize the interdependence between processes at different scales. The fractal alternative doesn't eliminate the insights gained through reductionist analysis but complements them with recognition of how processes at different scales influence and constrain each other — how the behavior of complex systems emerges from the interaction between patterns at multiple levels rather than being reducible to any single level of analysis.

This perspective aligns with complex systems approaches that emphasize non-linear dynamics, emergent properties, and the interaction between processes at different scales. It suggests that physical systems might be better understood through frameworks that capture their fractal organization across scales rather than focusing exclusively on their smallest components.

The Myth of Scale Separation

Another limitation of traditional approaches is the assumption of scale separation — the idea that processes at different scales can be treated independently, with each scale governed by its own laws and principles without fascinating influence from other scales. This assumption has led to the fragmentation of physics into separate domains — quantum mechanics for the very small, classical mechanics for the medium scale, relativity for the very large — each with its own mathematical framework and conceptual foundations.

Several problems with the scale separation assumption have become apparent:

- **Cross-scale feedback loops:** many physical systems develop feedback mechanisms that connect processes across different scales, creating integrated structures that can't be understood by treating each scale independently.
- **Scale-linking phenomena:** certain phenomena inherently connect processes across different scales, from quantum entanglement that links particles across space to phase transitions that coordinate behavior across multiple scales simultaneously.
- **Fractal boundaries:** the boundaries between different scales often exhibit fractal properties, creating transition regions where processes at different scales interact in complex ways rather than remaining neatly separated.
- **Information integration across scales:** complex systems often integrate information across different scales, creating capabilities that transcend what would be possible through processing at any single scale alone.

During my work with AI systems since founding zyrian.ai in 2013, I became interested in how these cross-scale interactions might inform more effective approaches to artificial intelligence. Traditional AI architectures often focus on processing at a single scale or use rigid hierarchical structures that don't capture the fluid, scale-linking dynamics of natural intelligence. The fractal perspective suggests that effective AI wouldn't just process information at multiple scales but would develop fluid, adaptive connections between these scales — connections that allow information and patterns to flow bidirectionally between different levels of processing rather than being confined to rigid hierarchical pathways.

This approach doesn't eliminate the value of scale-specific analysis but complements it with recognition of how scales interact and influence each other — how the behavior of complex systems emerges from the dynamic interaction between processes at multiple levels rather than being confined to any single scale of analysis.

Challenging the Observer-Reality Dichotomy

Perhaps the most fundamental limitation of traditional approaches is the sharp dichotomy between observers and the reality they observe — the idea that physical reality exists independently of observation, with clear boundaries between the observer, their measuring instruments, and the phenomena being studied. This dichotomy has been challenged by developments in quantum mechanics, where the act of observation appears to influence the behavior of the systems being observed in ways that can't be eliminated through more careful measurement techniques. Several problems with the observer-reality dichotomy have become apparent:

- **Measurement problem:** the role of measurement in quantum mechanics suggests a fundamental connection between observation and physical reality — a connection that can't be eliminated by treating the observer as separate from the system being observed.

- **Information and reality:** the increasing recognition that information might be as fundamental as matter and energy suggests that the observer's extraction and processing of information isn't separate from physical reality but an integral aspect of how reality manifests.
- **Consciousness and physical processes:** the relationship between consciousness and physical processes suggests that subjective experience might not be separate from objective reality but a different manifestation of the same underlying patterns.
- **Scale-linking observation:** the act of observation often connects processes across different scales, from the quantum interactions in measuring devices to the neural processes in the observer's brain to the social contexts in which observations are interpreted and communicated.

Have you ever wondered how your brain processes information across different scales simultaneously? This question lies at the heart of our exploration of fractal neural architectures. During my sabbatical in São Paulo, I had fascinating discussions with philosophers of science about these issues — about how different cultural and philosophical traditions approach the relationship between observers and the reality they observe.

What struck me was how certain non-Western traditions have long recognized the interdependence between observers and observed phenomena, offering perspectives that align in remarkable ways with the implications of quantum mechanics. The fractal perspective suggests that observers and the reality they observe might be understood not as separate domains but as different manifestations of the same underlying patterns — patterns that maintain certain essential properties while adapting to different contexts and scales. This doesn't eliminate the practical distinction between observers and observed phenomena but recognizes their deeper connection through shared fractal organization. This approach aligns with certain interpretations of quantum mechanics that emphasize the role of information and consciousness in physical reality.

It suggests that the relationship between observers and the systems they observe might be better understood through frameworks that capture their shared fractal organization rather than treating them as fundamentally separate domains.

FNA Approach to Physical Modeling

The fractal perspective on physics isn't just a theoretical framework but the foundation for practical approaches using Fractal Neural Architectures. Let's explore how FNAs might address aspects of physical modeling.

Multi-Scale Physical Simulation

Traditional approaches to physical simulation typically focus on a specific scale of analysis — either quantum, molecular, macroscopic, or cosmic — using different computational frameworks for each scale. This single-scale focus misses how physical processes emerge from the interaction between phenomena at different scales. FNAs address this limitation through multi-scale simulation that represents physical processes across different scales simultaneously:

- **Quantum-scale simulation:** representing subatomic particles and fields, including quantum superposition, entanglement, and field fluctuations.

- **Molecular-scale simulation:** capturing atomic and molecular interactions, including chemical bonding, molecular dynamics, and material properties.
- **Macroscopic-scale simulation:** modeling everyday objects and phenomena, including fluid dynamics, thermodynamics, and mechanical interactions.
- **Cosmic-scale simulation:** representing astronomical structures and processes, including gravitational interactions, stellar evolution, and galactic formation.

What makes this multi-scale approach particularly valuable is how it represents the connections between processes at different scales — how quantum phenomena influence molecular behavior, how molecular interactions create macroscopic properties, how macroscopic structures shape cosmic evolution, and how larger-scale patterns constrain smaller-scale possibilities. During my work with computational systems, I explored how these multi-scale models might be implemented using fractal neural architectures.

The key innovation was developing connectivity patterns that allowed information to flow naturally between different scales, creating integrated simulations that spanned from quantum to cosmic phenomena rather than treating each scale as a separate domain. This approach enables more sophisticated physical modeling — recognizing how processes at different scales influence and constrain each other, how properties emerge from cross-scale interactions, and how the overall fractal structure of physical reality evolves over time.

Fractal Uncertainty Modeling

Another key capability of FNAs for physical modeling is representing and leveraging the structured patterns in how uncertainty manifests across different scales — recognizing that unpredictability in physical systems isn't just random noise but exhibits fractal properties that can be characterized and utilized. This uncertainty modeling operates across several dimensions:

- **Quantum uncertainty characterization:** representing the structured patterns in quantum indeterminacy, including how uncertainty in particle position and momentum follows specific mathematical relationships rather than being purely random.
- **Scale-linking uncertainty propagation:** tracking how uncertainty at one scale influences and constrains uncertainty at other scales, capturing the complex ways in which unpredictability propagates across different levels of physical organization.
- **Pattern-based prediction under uncertainty:** instead of treating uncertainty as a limitation on prediction, the system leverages patterns in uncertainty itself to generate more robust predictions that account for the structured nature of unpredictability in physical systems.
- **Adaptive resolution adjustment:** based on patterns of uncertainty across scales, the system can adaptively adjust its resolution of analysis — focusing computational resources on regions and scales where uncertainty is most structured and consequential.

During my sabbatical, I had discussions with physicists about how these fractal approaches to uncertainty might transform our understanding of quantum phenomena. The key insight was that quantum uncertainty might not be mere randomness but might exhibit structured patterns across scales — patterns that could be leveraged to develop more effective approaches to quantum computation and simulation. This perspective suggests that uncertainty modeling wouldn't just characterize randomness but would recognize and leverage the fractal patterns in how uncertainty manifests across different scales and contexts — patterns that might reveal deeper structures in the nature of physical reality itself.

Temporal Fractal Analysis

Perhaps the most valuable application of FNAs to physics is temporal fractal analysis — identifying and characterizing the self-similar patterns that appear in physical processes across different time scales. Traditional time series analysis in physics typically focuses on specific temporal scales or uses hierarchical approaches that treat different time scales as separate domains. The fractal approach recognizes how temporal patterns often exhibit self-similarity across scales — how similar themes and structures can appear in processes unfolding across microseconds, seconds, hours, years, or eons. Several key capabilities emerge from this fractal approach:

- **Cross-scale temporal pattern identification:** the system can recognize similar patterns that appear across widely different time scales, identifying the fractal signature of particular physical processes.

- **Scale-linking temporal analysis:** rather than analyzing temporal patterns at a single time scale, the system can identify how patterns at different scales influence each other — how microsecond fluctuations connect to second-by-second dynamics, which connect to longer-term trends.

Have you noticed similar structures appearing in seemingly unrelated fields? This question lies at the heart of our exploration of fractal neural architectures.

- **Temporal emergence recognition:** the system can identify emergent temporal properties that aren't visible at any single time scale but emerge from the interaction between processes unfolding across different temporal horizons.
- **Adaptive temporal prediction:** based on fractal properties identified in past temporal patterns, the system can predict how processes are likely to evolve across different time scales, generating forecasts that integrate information from multiple temporal horizons.

During my early explorations with Excel simulations, I became fascinated by how simple recursive rules could generate complex temporal patterns that exhibited self-similarity across scales. These explorations suggested that temporal fractals might be fundamental to how physical processes unfold across different time scales — that the universe might be employing similar temporal principles throughout its diverse manifestations. This perspective suggests that temporal analysis wouldn't just focus on specific time scales or treat different scales as separate domains but would recognize and leverage the fractal patterns that integrate across different temporal horizons — patterns that might reveal deeper structures in how physical processes unfold through time.

In practice, implementations would include additional components like uncertainty quantification, adaptive resolution adjustment, and more sophisticated integration methods. The specific architecture would be adapted to the particular aspects of physical reality being modeled and the available computational resources.

Evaluation Metrics for Physical Models

Evaluating the quality of physical models requires metrics that go beyond simple accuracy measures to assess the fractal properties that contribute to effective representation of physical reality. Several specialized metrics have proven valuable:

- **Cross-scale coherence:** how consistently do model predictions maintain their essential qualities across different spatial and temporal scales? This measures the integration between processes at different levels of physical organization.
- **Emergent property prediction:** how effectively does the model predict properties that emerge from the interaction between processes at different scales? This assesses the model's capacity to capture phenomena that wouldn't be visible at any single scale of analysis.
- **Adaptive resolution efficiency:** how efficiently does the model allocate computational resources across different scales based on the fractal properties of the physical processes being modeled? This measures

the model's capacity to leverage fractal patterns for computational efficiency.

- **Uncertainty structure characterization:** how effectively does the model capture the structured patterns in how uncertainty manifests across different scales and contexts? This assesses the model's capacity to represent the fractal nature of unpredictability in physical systems.
- **Temporal pattern integration:** how well does the model integrate temporal patterns across different time scales? This measures the model's capacity to capture how physical processes unfold across multiple temporal horizons simultaneously.

These metrics provide a more comprehensive evaluation than simple accuracy or efficiency measures, helping identify specific strengths and vulnerabilities in physical models. During my work with physical modeling, I've found that qualitative assessment by domain experts can also provide valuable insights, particularly when guided by frameworks that help them recognize fractal patterns in physical processes. Experts can often identify recurring themes and patterns that might be missed by automated analysis, especially when given conceptual tools to recognize how these patterns manifest across different scales and contexts. Have you noticed similar structures appearing in seemingly unrelated fields?

This question lies at the heart of our exploration of fractal neural architectures. The fractal perspective suggests additional evaluation approaches focused on how models capture the scale-linking properties of physical systems. Rather than treating model quality as a single-scale property, we can assess how effectively models represent the integration between processes at different scales — how they capture the fractal organization that characterizes many physical phenomena.

Applications and Results

The theoretical advantages of Fractal Neural Architectures for physical modeling translate to practical benefits across various applications. Let's explore some specific examples and results.

Climate and Weather Modeling

One of the most promising applications of fractal approaches to physics is in climate and weather modeling — systems that predict how atmospheric and oceanic processes evolve across different spatial and temporal scales. Traditional climate and weather models typically focus on specific scales of analysis, using different approaches for short-term weather prediction and long-term climate modeling. This scale-specific focus misses how atmospheric and oceanic processes emerge from the interaction between phenomena at different scales — how microscale turbulence influences mesoscale weather patterns, which influence macroscale climate dynamics. Fractal approaches address this limitation through several innovations:

- **Multi-scale atmospheric representation:** rather than focusing on a single scale, fractal models represent atmospheric processes simultaneously across different scales, from local turbulence to global circulation patterns.
- **Scale-linking feedback integration:** the models capture how processes at different scales influence each other through feedback loops that span across multiple levels of atmospheric organization.

- **Fractal pattern leverage:** by recognizing self-similar patterns that appear across different scales in atmospheric data, the models achieve greater computational efficiency and predictive accuracy than approaches that treat each scale independently.
- **Temporal fractal analysis:** the models identify and leverage the self-similar patterns that appear in atmospheric processes across different time scales, from hourly fluctuations to seasonal cycles to multi-year trends.

During my visits to China, I had discussions with climate scientists about how these fractal approaches might transform weather and climate modeling. While I haven't directly implemented these ideas in operational forecasting systems, experimental applications have shown promising results. One particularly interesting application is in predicting extreme weather events that emerge from the interaction between processes at different scales — phenomena like hurricanes, tornadoes, and flash floods that traditional single-scale models often struggle to forecast accurately. By capturing how small-scale atmospheric instabilities can propagate and amplify across scales under certain conditions, fractal models offer new approaches to predicting these high-impact events. The fractal approach suggests that effective climate and weather modeling isn't just about increasing resolution or computational power but about better representing the scale-linking patterns that integrate across different aspects of atmospheric and oceanic dynamics.

Quantum Computing and Simulation

Another promising application area is quantum computing and simulation — systems that leverage quantum principles to perform computations or simulate quantum phenomena. Traditional approaches to quantum computing typically focus on manipulating quantum bits (qubits) through gate operations that transform their states according to the principles of quantum mechanics. While this approach has yielded crucial advances, it often struggles with the practical challenges of maintaining quantum coherence and scaling to larger systems. Fractal approaches offer several advantages for quantum computing and simulation:

- **Multi-scale quantum representation:** rather than focusing solely on qubit-level operations, fractal approaches represent quantum processes across different scales, from individual qubits to entangled subsystems to the entire quantum computer.
- **Fractal error correction:** by recognizing how errors propagate across different scales in quantum systems, fractal approaches enable more effective error correction strategies that leverage the structured patterns in how quantum noise manifests.
- **Scale-adaptive quantum algorithms:** fractal approaches enable algorithms that adaptively adjust their operation based on the specific fractal properties of the quantum system, allocating resources more effectively across different scales of quantum processing.
- **Temporal fractal quantum control:** by identifying and leveraging the self-similar patterns that appear in quantum dynamics across different time scales,

fractal approaches enable more effective control strategies that work with rather than against the natural temporal structure of quantum processes.

During my work with AI systems, I became interested in how these fractal approaches might transform quantum computing and simulation. The key insight was that effective quantum computation wouldn't just manipulate individual qubits but would leverage the scale-linking patterns that integrate across different levels of quantum organization. This perspective suggests that quantum advantage might come not just from the parallelism of superposition or the non-locality of entanglement but from the fractal organization that can emerge across different scales in quantum systems — organization that enables forms of computation that transcend what would be possible through single-scale processing alone.

Materials Science and Engineering

A third promising application area is materials science and engineering — the design and analysis of materials with specific properties and functions. Traditional approaches to materials science typically focus on specific scales of material organization — either atomic, molecular, microstructural, or macroscopic — using different techniques and frameworks for each scale. This scale-specific focus misses how material properties emerge from the interaction between structures at different scales. Fractal approaches offer several advantages for materials science and engineering:

- **Multi-scale material representation:** rather than focusing on a single scale, fractal approaches represent material structures simultaneously across different scales, from atomic arrangements to molecular configurations to microstructural patterns to macroscopic properties.
- **Emergent property prediction:** by capturing how processes at different scales influence each other, fractal approaches enable more accurate prediction of emergent material properties that arise from cross-scale interactions.
- **Fractal design optimization:** fractal approaches enable the design of materials with optimized properties across different scales, leveraging self-similar patterns to achieve specific functional characteristics.
- **Scale-linking material simulation:** by modeling how changes at one scale propagate to influence structures at other scales, fractal approaches enable more effective simulation of how materials respond to different conditions and treatments.

During my visits to oil prospection sites in China, I observed how these fractal approaches might transform materials analysis for resource extraction. The key insight was that effective material characterization wouldn't just analyze specific scales independently but would capture how structures at different scales integrate to determine overall material properties and behavior. This perspective suggests that materials science might advance not just through more detailed analysis at specific scales but through better understanding of how structures at different scales interact and influence each other — how the fractal organization of materials across scales determines their emergent properties and functions.

Comparison with Traditional Physical Modeling

Across these applications, several patterns emerge when comparing fractal approaches with traditional methods of physical modeling:

- **Emergent phenomenon prediction:** fractal approaches typically achieve more accurate prediction of emergent phenomena that arise from the interaction between processes at different scales, particularly for complex systems where single-scale analysis misses important cross-scale effects.
- **Computational efficiency:** by recognizing and leveraging self-similar patterns across scales, fractal approaches often achieve greater computational efficiency than methods that treat each scale independently, particularly for systems with fractal-like organization.
- **Uncertainty characterization:** fractal approaches typically provide more nuanced characterization of uncertainty in physical predictions, capturing the structured patterns in how unpredictability manifests across different scales rather than treating uncertainty as uniform randomness.
- **Adaptive resolution allocation:** fractal approaches enable more effective allocation of computational resources across different scales based on the specific properties of the physical systems being modeled, focusing resolution where it's most needed rather than applying uniform analysis across all scales.

These comparative advantages aren't marginal but often substantial, particularly for complex physical systems that span across different scales and exhibit emergent properties that wouldn't be visible at any single scale of analysis. During my three decades thinking about these issues and experimenting with different modeling approaches, I've become convinced that the fractal approach represents not just an incremental improvement but a fundamental reconceptualization of how we understand and model physical reality — one that better aligns with the complex, multi-scale nature of the universe itself.

Why It Matters

Transforming Our Understanding of Reality

The fractal approach to physics isn't just theoretically captivating but has profound implications for how we understand and interact with physical reality.

Enhancing Scientific Consciousness

Traditional approaches to physics often focus on specific domains or scales without developing broader awareness of the patterns that integrate across different aspects of physical reality. This fragmented approach can lead to specialized knowledge in isolated domains without transforming our overall understanding of the universe. The fractal perspective suggests a different path to scientific development — one centered on enhancing scientific consciousness:

- What if consciousness itself emerges from these fractal patterns? Consider how your own awareness integrates information across different time scales and sensory domains.

- **Pattern awareness across scales:** developing the ability to recognize physical patterns that span from quantum to cosmic scales, seeing how these patterns integrate across different levels of organization rather than treating each scale as a separate domain.
- **Self-similar theme recognition:** learning to identify recurring themes that appear in different aspects of physical reality, recognizing the fractal signature of particular physical principles across diverse manifestations.
- **Contextual pattern understanding:** developing awareness of how physical patterns adapt to different contexts while maintaining their essential qualities, distinguishing between surface variations and deeper pattern continuities.
- **Scale-linking dynamics recognition:** seeing how processes at different scales influence and constrain each other, recognizing the bidirectional flow of causality and information across different levels of physical organization.

During my sabbatical in São Paulo, I explored how these aspects of scientific consciousness might be cultivated through specific practices and frameworks. The key insight was that developing this consciousness isn't just about acquiring knowledge but about training perception — learning to see physical patterns that were previously invisible or only partially recognized. This enhanced scientific consciousness doesn't just improve specific theories or models but transforms the entire experience of engaging with physical reality — helping scientists and non-scientists alike see the physical world not as a collection of separate domains but as an integrated fractal pattern that unfolds across scales from quantum to cosmic, from microseconds to eons.

Implications for Scientific Development

The fractal perspective also has profound implications for how we approach scientific development — suggesting possibilities for growth that go beyond traditional models of specialization or interdisciplinary collaboration. Several alternative development paths emerge from the fractal approach:

- **Pattern evolution rather than domain expansion:** scientific development focuses not just on expanding knowledge within specific domains but on evolving the patterns that integrate across different fields and scales, creating more coherent understanding of physical reality.
- **Integrated growth across scales:** development occurs simultaneously across different scales of analysis, with insights at one scale informing and constraining theories at other scales through fractal propagation of patterns and principles.
- **Coherent differentiation:** scientific fields develop greater specialization while maintaining coherent connection through shared fractal patterns, creating more complex and resilient scientific understanding rather than merely balancing specialization and integration.

- **Emergent scientific creativity:** the scientific enterprise develops unique patterns that aren't reducible to individual disciplines but emerge from their specific interaction, creating possibilities that wouldn't be discovered through isolated disciplinary development.

During my early retirement and sabbatical, I've had time to observe how these development paths manifest in various scientific communities — how the most fruitful scientific advances often come not from isolated disciplinary progress but from recognizing patterns that integrate across different fields and scales. This perspective suggests that scientific development isn't about progressing toward a universal theory of everything but about evolving patterns that are uniquely adapted to the specific aspects of physical reality being studied — patterns that integrate insights from different scales and domains into coherent scientific structures that work for particular combinations of phenomena and contexts.

Ethical Considerations in Physical Modeling

The potential of fractal approaches to physical modeling brings significant ethical responsibilities, particularly as these approaches are implemented in technologies that analyze and influence the physical world. Several principles should guide the development of physical modeling technologies:

- **Environmental agency:** technologies should enhance rather than diminish the agency of natural systems, providing insights and options that respect the inherent patterns and processes of the physical world.

- **Diversity preservation:** development should respect and preserve the diversity of physical patterns rather than imposing standardized models based on particular theoretical or cultural frameworks.
- **Transparent analysis:** the basis for any physical analysis or intervention should be transparent and understandable to those affected, allowing for evaluation and potential override of algorithmic insights.
- **Scale-appropriate intervention:** interventions in physical systems should be appropriate to the specific scales and contexts involved, respecting the natural fractal organization of physical processes rather than imposing uniform approaches across different domains.
- **Balanced integration:** technologies should support balanced integration of human and natural systems rather than treating the physical world as merely a resource to be exploited or controlled.

During my work with AI systems since founding zyrian.ai in 2013, I've been particularly concerned with how these ethical considerations apply to technologies that analyze and influence physical systems. The most powerful physical technologies are those that enhance natural processes rather than replacing or controlling them — that augment the inherent capacity of physical systems for self-organization and adaptation rather than imposing external models. The fractal perspective suggests that these ethical considerations aren't separate from the technical development but integral to it — that the patterns of agency, diversity, transparency, scale-appropriateness, and integration we design into our systems will shape not just their performance but their impact on the physical world and human well-being.

Practical Framework

"Fractal Physics" Technique

To help you apply these concepts in your own exploration of physical reality, I offer this practical framework for fractal physics inspired by the principles discussed in this chapter.

Step 1

Recognize Physical Fractals

The foundation of fractal physics is recognizing the patterns that characterize physical processes across different scales. Practice identifying patterns across different domains and levels of organization:

- **Spatial patterns:** notice the recurring structures in physical systems across different spatial scales. How do branching patterns appear in river systems, plant structures, and neural networks? How do spiral patterns manifest in galaxies, weather systems, and seashells? What common principles might underlie these similar forms across different contexts?
- **Temporal patterns:** observe the rhythms and cycles that appear in physical processes across different time scales. How do oscillatory patterns manifest in phenomena from atomic vibrations to planetary orbits? How do growth and decay cycles appear across different temporal horizons? What self-similar structures can you identify in how processes unfold through time?
- **Organizational patterns:** identify the recurring structures in how physical systems organize them-

selves across different levels of complexity. How do similar network patterns appear in phenomena from molecular interactions to ecosystem relationships? How do hierarchical structures manifest across different domains? What common principles might underlie these organizational similarities?

- **Transformation patterns:** notice how physical systems change and evolve across different scales. How do similar patterns of growth, adaptation, and phase transition appear in phenomena from crystal formation to species evolution? What common principles might govern these transformation processes across different contexts?

During my visits to Niagara Falls and Foz do Iguaçu, I practiced this pattern recognition by observing the fractal structures in water flow — how similar patterns appeared at scales from tiny ripples to massive cascades. This observation suggested that fluid dynamics might exhibit fractal properties that connected phenomena across widely different scales, from microscopic movements to macroscopic patterns. This practice suggests that physical fractals are often more visible and pervasive than we realize — that with attention and practice, we can learn to recognize the self-similar patterns that appear across different aspects of physical reality.

Step 2

Identify Cross-Scale Connections

Once you've recognized patterns at different scales, look for connections between these scales — how processes at different levels influence and constrain each other:

- **Bottom-up emergence:** how do processes at smaller scales generate patterns and properties at larger scales? How do quantum fluctuations influence molecular behavior? How do molecular interactions create macroscopic properties? What principles govern how smaller-scale processes give rise to larger-scale phenomena?
- **Top-down constraint:** how do larger-scale patterns influence and constrain processes at smaller scales? How do environmental conditions shape molecular interactions? How do macroscopic structures influence quantum behavior? What principles govern how larger contexts shape smaller-scale dynamics?
- **Feedback loops across scales:** how do processes at different scales influence each other through feedback mechanisms? How do changes at one scale propagate to affect processes at other scales? What principles govern how these cross-scale feedback loops operate in different physical systems?
- **Information flow between scales:** how does information move between different scales in physical systems? How do patterns at one scale encode information that influences processes at other scales? What principles govern how information is transferred and transformed across different levels of physical organization?

During my early explorations with Excel simulations, I became fascinated by how simple recursive rules could generate complex structures that exhibited connections across scales. These explorations suggested that cross-scale connections might be fundamental to how physical reality organizes itself — that the universe might be employing similar principles of scale integration throughout its diverse manifestations. This practice suggests that cross-scale connections aren't just theoretical constructs but observable aspects of physical reality — that with attention and appropriate tools, we can identify and characterize how processes at different scales influence and constrain each other through various mechanisms of connection and feedback.

Step 3

Analyze Fractal Properties

The next step is analyzing the specific fractal properties of physical systems — the mathematical characteristics that describe how patterns manifest and propagate across scales:

- **Self-similarity analysis:** to what degree do patterns maintain their essential structure across different scales? Is the self-similarity exact (as in mathematical fractals) or statistical (as in many natural systems)? How does the nature of self-similarity vary across different aspects of the system?
- **Fractal dimension calculation:** what is the fractal dimension of different aspects of the system? How does this dimension compare to the topological dimension of the space in which the system exists? What does the specific fractal dimension reveal about how the system fills space or unfolds through time?
- **Scale invariance assessment:** which properties remain invariant across changes in scale, and which transform in specific ways? What mathematical relationships describe how different aspects of the system change

with scale? What do these scaling relationships reveal about the underlying principles governing the system?

- **Boundary complexity characterization:** how complex are the boundaries between different regions or states in the system? How does this complexity manifest across different scales? What does the specific nature of boundary complexity reveal about the processes that generate and maintain these boundaries?

During my work with physical systems, I've found that these fractal analyses aren't just abstract mathematical exercises but practical tools for understanding how physical reality organizes itself across scales. The specific fractal properties of a system often reveal deeper principles that might not be visible through other forms of analysis. This practice suggests that fractal analysis isn't just for specialized mathematicians but can be accessible to anyone with appropriate conceptual frameworks and analytical tools — that with practice, we can learn to characterize and interpret the fractal properties of physical systems across different domains and scales.

Step 4

Apply Fractal Thinking to Physical Problems

The final step is applying fractal thinking to specific physical problems or questions — using the patterns, connections, and properties you've identified to develop new insights or approaches:

- **Identify scale-linking aspects:** for the specific problem you're addressing, identify which aspects involve connections between processes at different scales. These scale-linking aspects often provide leverage points where fractal approaches can offer particular value.

- **Look for self-similar patterns:** examine whether similar patterns appear across different scales relevant to your problem. These self-similar patterns might reveal underlying principles that could inform more effective approaches.
- **Consider cross-scale effects:** analyze how changes at one scale might propagate to influence processes at other scales. These cross-scale effects often explain phenomena that would be puzzling when viewed from a single-scale perspective.
- **Leverage fractal properties:** design approaches that work with rather than against the fractal properties of the physical systems involved. These fractal-aligned approaches often achieve greater effectiveness and efficiency than methods that ignore or fight against natural patterns.
- **Integrate across scales:** develop solutions that integrate across different scales rather than focusing exclusively on a single level of analysis. These integrated approaches often address root causes rather than merely treating symptoms at a particular scale.

During my three decades thinking about these issues, I've observed how this fractal thinking has transformed approaches to various physical problems — not through imposing abstract mathematical frameworks but through recognizing and working with the natural patterns that integrate across different scales in physical reality. The key insight is that physical systems often exhibit fractal properties that can be leveraged rather than ignored or fought against — that solutions aligned with the natural fractal organization of physical reality typically achieve greater effectiveness and sustainability than approaches that focus exclusively on a single scale or domain.

Call to Action

Exploring Your Fractal Universe

As we conclude this chapter on the laws of physics, I invite you to begin applying these insights to enhance your own understanding of physical reality. Here are specific actions you can take: practice fractal observation in your physical environment. Set aside time regularly to notice patterns that appear across different scales in natural and built systems around you. Keep a fractal journal where you record these observations, looking for self-similar themes that recur across different aspects of physical reality.

Experiment with simple fractal simulations using accessible tools like spreadsheets or basic programming. Create models that apply simple rules recursively across scales and observe the complex patterns that emerge. Notice how these patterns compare to those you observe in natural systems. Develop cross-scale awareness in your interaction with physical systems. Practice maintaining awareness of how processes at different scales influence and constrain each other, seeing both immediate interactions and their connection to larger patterns across scales.

Initiate fractal conversations with friends or colleagues interested in science and nature.

Share the concept of physical fractals and explore together what patterns you've noticed across different domains and scales. These conversations can reveal connections and insights that might not be visible through individual observation alone. The fractal nature of physical laws isn't just a theoretical framework but a practical approach to understanding and engaging with the universe around us. By recognizing and working with the fractal patterns in physical reality, you can develop a more integrated understanding of how the universe organizes itself across scales from quantum to cosmic, from microseconds to eons.

In the next chapter, we'll explore how fractal principles apply to universal patterns — how the fundamental structures we've examined in neural architectures, language, consciousness, relationships, and physical laws might be understood as different manifestations of deeper patterns that transcend specific domains and contexts.

CHAPTER 9

Universal Patterns

Narrative Opening: The Pattern That Connects

The conference room fell silent as Dr. Elena Sato concluded her presentation. On the large display behind her, a series of visualizations showed strikingly similar patterns across wildly different domains — neural networks, language structures, conscious awareness, human relationships, and physical laws. "What we're proposing", she said, looking across the diverse audience of scientists, philosophers, and technologists, "is that these aren't just superficial similarities. They represent a fundamental pattern that connects across domains — a universal template that manifests in different forms depending on context."

Professor Chen nodded thoughtfully. "You're suggesting that fractal organization isn't just a mathematical curiosity or a specific feature of certain systems, but a universal principle — perhaps the universal principle — underlying complexity across all domains." "Exactly", Elena confirmed. "Whether we're examining how neurons organize information, how languages structure meaning, how consciousness integrates experience, how relationships evolve over time, or how physical laws manifest across scales — we keep encountering the same essential pattern: self-similar structures that maintain their core properties while adapting to different contexts and scales." From the back of the room, a skeptical voice rose. “But isn’t this just pattern-matching? Humans are notorious for seeing patterns even where none exist.”

Kai Mercer stepped forward to address the question. “That’s a valid concern. We’ve been careful to distinguish between superficial similarities and deeper structural homologies. What we’re identifying aren’t just visual resemblances but functional parallels — similar organizational principles that solve similar problems across different domains.” He gestured toward the display, which now showed mathematical representations of the patterns they’d identified. “These aren’t just qualitative observations but quantifiable relationships that can be formally modeled and tested.” “But what’s the mechanism?” another audience member asked. “How would the same pattern emerge independently across such different domains?”

Elena smiled. “That’s where it gets interesting. We don’t believe these patterns emerged independently. We’re proposing that they represent different manifestations of a more fundamental principle — a universal template for how complex systems organize themselves across scales.” “Are you suggesting some kind of mystical unity?” someone asked skeptically. “Not at all”, Kai replied. “We’re suggesting that certain organizational principles are mathematically optimal for specific functions — like information processing, adaptation to changing environments, or balancing stability and flexibility. Natural selection, broadly understood, would favor these optimal patterns across different domains.”

Professor Chen added, “Think of it as convergent evolution at the level of organizational principles. Just as similar body shapes evolved independently in fish and dolphins because they’re optimal for swimming, similar organizational principles might emerge across different domains because they’re optimal for functions like information processing or adaptive learning." Elena nodded. "But there's another possibility that's even more intriguing. What if these patterns don't just represent similar solutions to similar problems, but different manifestations of the same underlying reality? What if the fractal patterns we observe in neural networks, language, consciousness, relationships, and physical laws are different expressions of a more fundamental pattern that transcends specific domains?" The room grew quiet as the implications of this suggestion sank in. It wasn't just a theory about similar patterns across domains but a radical reconceptualization of reality itself — a perspective that suggested the boundaries between different domains might be more permeable than traditionally assumed, that the same fundamental principles might manifest across what we've traditionally treated as separate realms of existence.

What if the same fundamental patterns underlie everything from neurons to galaxies?

This question lies at the heart of our exploration of fractal neural architectures. "If you're right", Professor Chen said slowly, "this wouldn't just change how we understand specific systems but how we understand reality itself. It would suggest that the fragmentation of knowledge into separate disciplines might be missing something essential — that there might be a pattern that connects across what we've traditionally treated as separate domains." Elena nodded. "Exactly. And that pattern that connects might be the key to developing more integrated approaches across disciplines — approaches that recognize and leverage the universal templates that manifest in different forms across neural networks, language, consciousness, relationships, physical laws, and perhaps domains we haven't even considered yet."

As the discussion continued, the boundaries between different fields seemed to dissolve, revealing glimpses of a more integrated understanding — one where the same fundamental patterns appeared across different manifestations, creating a universe that was both endlessly diverse and profoundly unified.

The Pattern That Connects

The narrative above introduces a profound possibility: that the fractal patterns we've explored in previous chapters might not be isolated phenomena but different manifestations of a more fundamental pattern that connects across domains. This perspective challenges traditional boundaries between disciplines while opening new avenues for integrated understanding.

Cross-Domain Fractal Patterns

One of the most striking aspects of fractal patterns is how similar structures appear across widely different domains, from neural networks to language to consciousness to relationships to physical laws. In neural networks, we observe fractal patterns in how information is organized and processed across different scales — from individual neurons to local circuits to regional networks to global brain organization. These patterns create systems that can integrate information across scales while maintaining both specialization and coordination. In language, we find fractal patterns in how meaning is structured across different levels — from phonemes to morphemes to words to sentences to narratives.

These patterns enable language to convey complex meanings through the recursive combination of simpler elements, creating systems that are both rule-governed and creatively open. In consciousness, we observe fractal patterns in how awareness integrates information across different scales — from momentary sensations to extended experiences to lifetime narratives. These patterns create subjective experiences that maintain coherence across scales while continuously evolving in response to new inputs. In relationships, we find fractal patterns in how connection unfolds across different time scales — from momentary interactions to conversations to daily rhythms to lifetime bonds. These patterns enable relationships to maintain their essential qualities across contexts while adapting to changing circumstances and needs.

In physical laws, we observe fractal patterns in how similar principles manifest across different scales — from quantum to molecular to macroscopic to cosmic. These patterns create a physical reality that exhibits both consistency and diversity across scales, with similar themes appearing in different contexts throughout the universe. What makes these cross-domain patterns particularly interesting is how they suggest deeper principles that transcend specific contexts — principles that might reflect fundamental aspects of how complex systems organize themselves across scales regardless of their specific domain or content. During my sabbatical in São Paulo, I spent countless hours contemplating these cross-domain patterns, often while observing natural systems like the branching patterns of trees or the flow patterns in rivers.

What struck me was how the same essential structures appeared across domains that we typically treat as separate — how similar organizational principles seemed to be operating in systems as diverse as neural networks, languages, ecosystems, and physical processes. This observation led me to wonder whether these similarities might reflect more than just superficial resemblances — whether they might represent different manifestations of more fundamental patterns that transcend specific domains. Just as the Sierpiński triangle can be generated through different processes that converge on the same fractal structure, perhaps the fractal patterns we observe across different domains represent different paths to similar organizational principles. This perspective offers a new way to understand both the diversity and unity of complex systems — how they can manifest in dramatically different forms while maintaining underlying connections, how patterns in one domain might inform our understanding of seemingly unrelated domains.

Functional Parallels Across Domains

Beyond structural similarities, fractal patterns across different domains exhibit striking functional parallels — similar organizational principles that solve similar problems across diverse contexts. Several key functional parallels emerge across domains:

- **Multi-scale information integration:** across neural networks, language, consciousness, relationships, and physical systems, fractal patterns enable the integration of information across different scales. This multi-scale integration creates capabilities that wouldn't be possible through single-scale processing alone, allowing systems to maintain both local specialization and global coherence.
- **Adaptive flexibility within constraints:** fractal patterns across domains enable systems to adapt to changing circumstances while maintaining their es-

sential structure. This balance between flexibility and constraint allows for both stability and innovation, creating systems that can evolve without losing their core organization.

- **Efficient resource utilization:** across different domains, fractal patterns optimize the use of limited resources — whether physical materials, energy, computational capacity, or attention. This efficiency emerges from how fractal structures distribute resources across scales, maximizing functional capacity while minimizing costs.
- **Emergent complexity from simple principles:** fractal patterns across domains enable complex structures and functions to emerge from relatively simple principles applied recursively. This emergent complexity allows sophisticated capabilities to develop without requiring explicit design or control at each level of organization.

These functional parallels suggest that fractal organization might represent an optimal solution to certain fundamental challenges that complex systems face across different domains — challenges like integrating information across scales, adapting to changing environments, utilizing limited resources efficiently, and generating complexity from simplicity. During my work with AI systems since founding zyrian.ai in 2013, I became increasingly aware of how these functional parallels might inform more effective approaches to artificial intelligence. If natural systems across different domains converge on similar fractal solutions to similar functional challenges, perhaps artificial systems should leverage these same organizational principles rather than imposing rigid structures that might miss the essential flexibility and efficiency of natural systems.

This perspective suggests that the most effective artificial systems wouldn't just mimic specific aspects of natural intelligence but would leverage the same fractal principles that appear across different natural systems — principles that might represent optimal solutions to the fundamental challenges of complex information processing and adaptive learning.

Mathematical Homologies

Beyond structural similarities and functional parallels, fractal patterns across different domains exhibit mathematical homologies — formal relationships that can be described through similar mathematical frameworks despite appearing in diverse contexts. Several key mathematical homologies appear across domains:

- **Power law distributions:** across neural networks, language, social relationships, and physical systems, we observe similar power law distributions — patterns where the frequency of an element is inversely proportional to its size or intensity raised to a constant power. These distributions create scale-free networks with similar mathematical properties across different domains.
- **Self-similar scaling relationships:** fractal patterns across domains exhibit similar scaling relationships — mathematical functions that describe how properties change with scale. These relationships often follow similar mathematical forms despite appearing in widely different contexts, suggesting deeper connections in how complex systems organize themselves across scales.

- **Fractal dimensions:** the fractal dimensions of patterns across different domains often fall within similar ranges, creating structures with comparable space-filling properties despite their diverse manifestations. These similar dimensions suggest constraints on how complex systems can effectively organize themselves across scales.
- **Information-theoretic relationships:** fractal patterns across domains exhibit similar information-theoretic properties — relationships between entropy, complexity, and information transfer that follow comparable mathematical patterns despite appearing in different contexts.

These mathematical homologies suggest that the similarities between fractal patterns across domains aren't just qualitative observations but quantifiable relationships that can be formally modeled and analyzed. They point toward deeper mathematical principles that might underlie how complex systems organize themselves across scales regardless of their specific domain or content. During my visits to China, particularly in remote regions near the Kazakhstan frontier, I had fascinating discussions with mathematicians who were exploring these cross-domain homologies from different theoretical perspectives.

What struck me was how these mathematical relationships transcended traditional boundaries between disciplines — how they revealed connections between systems that we typically treat as belonging to separate domains of inquiry. This perspective suggests that mathematics might provide a universal language for understanding the patterns that connect across different domains — not just as a tool for modeling specific systems but as a framework for recognizing deeper connections between systems that might appear unrelated on the surface.

Evolutionary Convergence or Unified Reality?

The cross-domain patterns, functional parallels, and mathematical homologies we've explored raise a profound question: Do these similarities represent evolutionary convergence toward optimal solutions, or do they reflect different manifestations of a more unified underlying reality? The evolutionary convergence hypothesis suggests that similar patterns emerge independently across different domains because they represent optimal solutions to similar functional challenges. Just as similar body shapes evolved independently in fish and dolphins because they're optimal for swimming, similar organizational principles might emerge across different domains because they're optimal for functions like information processing or adaptive learning.

Several factors support this hypothesis:

- **Functional optimization:** fractal patterns often represent mathematically optimal solutions to challenges like maximizing surface area within limited volume, efficiently distributing resources across scales, or balancing competing functional requirements.
- **Selection pressures:** across different domains, systems face similar selection pressures for efficiency, adaptability, and robustness, which might favor similar organizational principles regardless of the specific context.

- **Convergent evolution:** the history of evolution shows numerous examples of convergence toward similar structures in unrelated lineages, suggesting that certain organizational principles might be discovered independently across different domains.

The unified reality hypothesis, by contrast, suggests that similar patterns across domains might reflect different manifestations of the same underlying reality — that the boundaries between what we treat as separate domains might be more permeable than traditionally assumed, with the same fundamental patterns manifesting in different forms across neural networks, language, consciousness, relationships, and physical laws.

Several factors support this hypothesis:

- **Scale-linking phenomena:** many phenomena inherently connect processes across different domains, from how neural patterns enable conscious experience to how language shapes social relationships to how physical laws constrain biological possibilities.
- **Boundary transcendence:** the boundaries between traditional domains often blur upon closer examination, revealing phenomena that transcend conventional categories and suggest deeper connections between what we typically treat as separate realms.
- **Unified mathematical descriptions:** similar mathematical frameworks can often describe patterns across different domains, suggesting that these domains might represent different manifestations of the same underlying mathematical principles.

During my early retirement and sabbatical, I've had the opportunity to build a robust infrastructure of AI "agents" to collaborate with me on various projects. These collaborations have provided a laboratory for exploring how principles from different domains might be integrated into more effective technological approaches — how insights from neuroscience, linguistics, psychology, sociology, and physics might inform more integrated approaches to artificial intelligence.

Personal Journey

Discovering Universal Patterns

My path to understanding universal patterns wasn't through traditional academic research but through personal experiences and observations that gradually coalesced into a coherent perspective.

Learning from Cross-Cultural Perspectives

A third strand in my exploration of universal patterns came through cross-cultural experiences — seeing how different cultural traditions often recognized similar patterns across domains but conceptualized and expressed these patterns through different frameworks and metaphors. As someone who immigrated from Brazil to the United States and later traveled extensively in places like China, I've had the opportunity to engage with diverse cultural perspectives on the patterns that connect across different domains of experience and knowledge.

What struck me was how certain traditional knowledge systems often recognized connections across domains that

Western scientific disciplines typically treat as separate — how traditional Chinese medicine connected patterns across physical, emotional, and social dimensions, how indigenous knowledge systems integrated observations across ecological, social, and spiritual domains, how ancient philosophical traditions recognized parallels between patterns in nature and patterns in human experience. These cross-cultural perspectives suggested that the fragmentation of knowledge into separate disciplines might be a relatively recent and culturally specific approach rather than a necessary feature of human understanding. They raised the possibility that more integrated perspectives might recover insights that have been obscured by the specialization and fragmentation of knowledge in modern academic disciplines.

During my visits to remote regions of China near the Kazakhstan frontier, I had fascinating conversations with scholars who were working to integrate traditional knowledge systems with contemporary scientific frameworks. What emerged from these discussions wasn't a rejection of scientific rigor but an expansion of what counts as relevant evidence and meaningful connection — an approach that could recognize patterns across domains that might be missed when each domain is treated as a separate conceptual universe.

The Emergence of Integrated Understanding

Perhaps the most valuable insights from my exploration of universal patterns came not from any single observation or experience but from the gradual emergence of a more integrated understanding — a perspective that could recognize connections across domains without collapsing important distinctions or imposing artificial unity.

Like many people exploring unconventional ideas, I've experienced both the excitement of recognizing patterns across domains and the challenge of distinguishing meaningful connections from superficial similarities. What gradually became clear through this process was that understanding universal patterns wasn't about finding a single unified theory that would reduce all phenomena to the same framework but about recognizing how similar organizational principles might manifest differently across diverse contexts. The most valuable lessons came from recognizing how these patterns operated across different domains — how similar mathematical structures could appear in contexts as diverse as neural networks, language systems, conscious experience, social relationships, and physical processes, yet with crucial differences that reflected the specific conditions of each domain.

During my three decades thinking about these issues, I've become increasingly convinced that recognizing universal patterns offers a path beyond both rigid reductionism and chaotic relativism. It suggests that while the universe exhibits remarkable diversity across different domains and scales, this diversity emerges from deeper patterns that maintain certain essential properties while adapting to specific conditions. This perspective has informed not just my theoretical understanding but my practical approaches to technology development, relationship building, and personal growth — creating frameworks that can recognize and leverage the patterns that connect across different domains of experience and knowledge.

Contrarian Perspective

Beyond Disciplinary Fragmentation

The universal pattern approach challenges several fundamental assumptions in traditional academic and intellectual frameworks. Let me explain why I believe some established approaches miss the connections that integrate across different domains of knowledge and experience. Do you see how these patterns repeat across different domains? This question lies at the heart of our exploration of fractal neural architectures.

Why Disciplinary Specialization Falls Short

Traditional approaches to knowledge often rely on disciplinary specialization — the division of inquiry into separate fields like physics, biology, psychology, linguistics, and sociology, each with its own concepts, methods, and standards of evidence. While this approach has yielded remarkable insights within specific domains, it often misses the patterns that connect across different fields. One of my most profound moments of cross-domain pattern recognition came during an interdisciplinary conference where I was presenting on neural networks. A botanist approached me afterward, pointing out how the activation patterns in my neural models resembled the growth patterns of certain plants. This chance conversation sparked a collaboration that revealed striking similarities between biological growth algorithms and neural learning processes — a connection I might never have made without that serendipitous exchange.

Several limitations of disciplinary specialization have become increasingly apparent:

- **Boundary blindness:** specialized disciplines often develop blind spots at their boundaries, missing phenomena that don't fit neatly within established categories or that emerge from the interaction between processes typically studied by different fields.
- **Conceptual isolation:** disciplines typically develop their own specialized concepts and terminology, creating artificial barriers to recognizing similar patterns that might be described through different language in different fields.
- **Methodological constraints:** each discipline typically develops its own methodological approaches, limiting the ability to recognize patterns that might require integrated methods spanning across traditional disciplinary boundaries.
- **Fragmented understanding:** perhaps most fundamentally, disciplinary specialization fragments our understanding of reality into separate domains, missing the patterns that integrate across different aspects of experience and knowledge.

During my travels and cross-cultural experiences, I've observed how different intellectual traditions approach these limitations in different ways. Western academia has traditionally emphasized disciplinary specialization, while many Eastern and indigenous traditions have maintained more integrated approaches that recognize connections across domains that Western frameworks treat as separate.

The universal pattern alternative doesn't eliminate the value of specialized knowledge but complements it with recognition of the patterns that connect across different domains — patterns that might be missed when each field is treated as a separate conceptual universe. It suggests that the most complete understanding comes not from reducing all phenomena to a single framework but from recognizing how similar patterns manifest differently across diverse contexts. This perspective aligns with emerging transdisciplinary approaches that seek to transcend traditional boundaries between disciplines. It suggests that knowledge might be better organized around patterns that appear across different domains rather than being fragmented into separate fields based on the specific content being studied.

The Myth of Domain Independence

Another limitation of traditional approaches is the assumption of domain independence — the idea that different fields study fundamentally separate aspects of reality that can be understood in isolation from each other. This assumption has led to the fragmentation of knowledge into separate disciplines, each with its own concepts, methods, and standards of evidence. Several problems with the domain independence assumption have become apparent:

- **Interconnected reality:** the aspects of reality studied by different disciplines aren't truly independent but deeply interconnected — neural processes enable consciousness, which shapes language, which influences relationships, which exist within physical reality.
- **Boundary phenomena:** many of the most intriguing and essential phenomena emerge at the boundaries

between traditional domains — phenomena that can't be fully understood through the concepts and methods of any single discipline.

- **Common patterns across domains:** similar patterns often appear across domains that are traditionally treated as separate, suggesting deeper connections that transcend conventional disciplinary boundaries.
- **Integrated experience:** human experience integrates across domains that academic frameworks treat as separate, creating a gap between lived reality and fragmented disciplinary knowledge.

During my work with AI systems, I became interested in how these cross-domain connections might inform more effective approaches to artificial intelligence. Traditional AI architectures often focus on specific domains or capabilities, missing the integrated nature of natural intelligence that spans across functions like perception, language, reasoning, emotion, and social interaction. The universal pattern perspective suggests that effective AI wouldn't just develop separate capabilities in different domains but would recognize and leverage the patterns that integrate across different aspects of intelligence — patterns that might enable more coherent and adaptive functioning than approaches that treat each capability as a separate module.

This approach doesn't eliminate the value of domain-specific analysis but complements it with recognition of how domains interact and influence each other — how the patterns in one area connect to and constrain possibilities in others, creating an integrated reality that transcends traditional disciplinary boundaries.

Challenging the Theory-Practice Dichotomy

Perhaps the most fundamental limitation of traditional approaches is the sharp dichotomy between theory and practice — the idea that abstract knowledge exists separately from practical application, with clear boundaries between theoretical understanding and practical engagement with the world. This dichotomy has led to the separation of academic knowledge from practical wisdom, creating artificial barriers between those who develop theoretical frameworks and those who apply knowledge in real-world contexts.

Several problems with the theory-practice dichotomy have become apparent:

- **Embodied knowledge:** much of our most valuable knowledge isn't abstract theory but embodied understanding that emerges from practical engagement with the world — knowledge that can't be fully captured through theoretical frameworks alone.
- **Practice-theory feedback:** theoretical understanding both informs and is informed by practical application, creating feedback loops that transcend the conventional boundary between theory and practice.
- **Pattern recognition across contexts:** some of the most valuable insights come from recognizing patterns across different contexts of practical engagement — patterns that might not be visible through theoretical analysis alone.
- **Integrated knowing-doing:** the most effective approaches often integrate knowing and doing rather

> than treating them as separate domains — approaches that recognize how understanding emerges through active engagement rather than passive observation.

During my sabbatical in São Paulo, I had fascinating discussions with practitioners across different fields about how they integrated theoretical understanding with practical wisdom. What struck me was how the most effective approaches didn't treat theory and practice as separate domains but recognized the patterns that connected across different forms of knowing and doing. The universal pattern perspective suggests that knowledge isn't just abstract theory but emerges from the recognition of patterns across different contexts of engagement with the world.

It proposes that the most complete understanding comes not from separating theory from practice but from recognizing how similar patterns manifest across both theoretical frameworks and practical applications. This approach doesn't eliminate the distinction between different forms of knowledge but recognizes how they might be connected through common patterns that transcend traditional boundaries between theory and practice, abstract and concrete, knowing and doing.

Beyond the Objective-Subjective Dichotomy

A fourth limitation of traditional approaches is the sharp dichotomy between objective and subjective — the idea that there's a clear boundary between objective facts about the world and subjective experiences of individuals, with the former belonging to science and the latter to art, literature, or personal reflection. This dichotomy has led to the separation of objective knowledge from subjective experience, creating artificial barriers between disciplines that focus on external reality and those that explore internal experience.

Several problems with the objective-subjective dichotomy have become apparent:

- **Observer-dependent reality:** developments in fields from quantum physics to cognitive science suggest that what we perceive as objective reality is shaped by the nature of observation itself — that the boundary between observer and observed isn't as clear as traditionally assumed.
- **Intersubjective knowledge:** much of our most valuable knowledge exists in the intersubjective space between purely objective facts and purely subjective experiences — knowledge that emerges from shared patterns of understanding across different perspectives.
- **Pattern recognition across domains:** similar patterns often appear across domains traditionally categorized as objective or subjective, suggesting deeper connections that transcend this conventional boundary.
- **Integrated experience:** human experience integrates across aspects traditionally categorized as objective or subjective, creating a gap between lived reality and frameworks that maintain a sharp boundary between these domains.

During my early retirement and sabbatical, I've had time to explore these issues through both theoretical reflection and practical observation. What gradually became clear was how the universe might transcend the objective-subjective dichotomy through patterns that appear across both external reality and internal experience — patterns that suggest deeper connections between what we traditionally treat as separate domains.

The universal pattern perspective suggests that similar organizational principles might appear across both objective and subjective domains — that the patterns we observe in external reality might have parallels in the structures of subjective experience, creating connections that transcend the conventional boundary between objective and subjective. This approach doesn't eliminate the distinction between different aspects of experience but recognizes how they might be connected through common patterns that transcend traditional boundaries between objective and subjective, external and internal, scientific and humanistic.

FNA Approach to Universal Patterns

The universal pattern perspective isn't just a theoretical framework but the foundation for practical approaches using Fractal Neural Architectures. Let's explore how FNAs might address aspects of cross-domain pattern recognition.

Universal Pattern Templates

Traditional approaches to pattern recognition typically focus on specific domains — either neural, linguistic, social, or physical — using different computational frameworks for each domain. This single-domain focus misses how similar patterns might appear across different fields and contexts. FNAs address this limitation through universal pattern templates that can recognize and leverage similarities across different domains:

- **Multi-domain pattern representation:** representing patterns across different domains simultaneously, from neural networks to language to social relationships to physical systems.
- **Cross-domain similarity identification:** recognizing similarities between patterns in different domains based on structural, functional, temporal, and informational parallels.
- **Abstract template extraction:** generating abstract templates that capture the essential structure of patterns across different manifestations, creating flexible representations that can be applied across diverse contexts.
- **Pattern-based cross-domain transfer:** applying patterns recognized in one domain to facilitate understanding and prediction in other domains, leveraging similarities to accelerate learning and innovation across fields.

What makes this universal template approach particularly valuable is how it represents the connections between patterns across different domains — how neural patterns relate to linguistic structures, how linguistic structures shape social relationships, how social patterns reflect physical principles, and how physical patterns constrain neural possibilities. During my work with pattern recognition systems, I explored how these universal templates might be implemented using fractal neural architectures. The key innovation was developing connectivity patterns that allowed information to flow naturally between different domains, creating integrated representations that spanned across neural, linguistic, social, and physical patterns rather than treating each domain as a separate category.

This approach enables more sophisticated pattern recognition — identifying similarities across domains that might be missed when each field is analyzed in isolation, recognizing how patterns in one area might inform understanding in others, and developing more integrated approaches to complex challenges that span traditional disciplinary boundaries. Have you ever wondered how your brain processes information across different scales simultaneously? This question lies at the heart of our exploration of fractal neural architectures. Several key capabilities emerge from this universal template approach:

- **Abstract pattern representation:** the system can represent patterns at different levels of abstraction, from specific manifestations in particular domains to more general templates that capture similarities across diverse contexts.
- **Flexible pattern matching:** rather than requiring exact matches to rigid templates, the system can recognize flexible variations on core patterns — identifying how similar organizational principles might manifest differently across different domains and contexts.
- **Hierarchical pattern organization:** the system can organize patterns hierarchically, from more specific templates for particular domains to more abstract templates that capture similarities across diverse contexts.
- **Evolving pattern templates:** through exposure to diverse manifestations across different domains, pattern templates can evolve to become more abstract and flexible while maintaining their essential structure.

During my three decades thinking about these issues, I've become increasingly convinced that universal pattern templates offer a powerful approach to understanding the connections across different domains of knowledge and experience. These templates wouldn't reduce all phenomena to a single framework but would recognize how similar organizational principles might manifest differently across diverse contexts. This perspective suggests that some of the most valuable insights come from recognizing the patterns that connect across different domains — from seeing how similar structures might appear in neural networks, language systems, social relationships, and physical processes, revealing deeper principles that transcend specific content or context.

In practice, implementations would include additional components like uncertainty quantification, hierarchical pattern organization, and more sophisticated similarity metrics. The specific architecture would be adapted to the particular domains being integrated and the available data sources.

Evaluation Metrics for Universal Patterns

Evaluating the quality of universal pattern recognition requires metrics that go beyond domain-specific accuracy to assess how effectively patterns are recognized and applied across different contexts. Several specialized metrics have proven valuable:

- **Cross-domain transfer accuracy:** how effectively can patterns recognized in one domain be applied to facilitate recognition or prediction in other domains? This measures the system's ability to leverage similarities across contexts.
- **Abstract template quality:** how well do abstract templates capture the essential structure of patterns

while allowing for variation in specific manifestations? This assesses the balance between specificity and generality in pattern representations.

- **Integration coherence:** how coherently does the system integrate patterns across different domains? This measures whether cross-domain connections create a coherent understanding or merely a collection of separate observations.
- **Creative recombination value:** how valuable are the novel combinations generated by applying patterns from one domain to others? This assesses the system's ability to generate insights through cross-domain application.
- **Boundary phenomenon capture:** how effectively does the system represent phenomena that emerge at the boundaries between traditional domains? This measures the ability to capture processes that wouldn't be visible through domain-specific analysis alone.

These metrics provide a more comprehensive evaluation than domain-specific accuracy measures, helping identify specific strengths and vulnerabilities in cross-domain pattern recognition. During my work with pattern recognition systems, I've found that qualitative assessment by domain experts can also provide valuable insights, particularly when experts from different fields evaluate the same patterns from their respective perspectives. This multi-perspective evaluation can reveal aspects of patterns that might not be captured by any single disciplinary framework. The universal pattern perspective suggests additional evaluation approaches focused on how effectively systems capture the connections across different domains.

Rather than treating pattern quality as a domain-specific property, we can assess how patterns facilitate understanding and prediction across traditional boundaries — how they reveal connections that might be missed when each domain is analyzed in isolation.

Applications and Results

The theoretical advantages of Fractal Neural Architectures for universal pattern recognition translate to practical benefits across various applications. Let's explore some specific examples and results.

Interdisciplinary Research and Innovation

One of the most promising applications of universal pattern approaches is in interdisciplinary research and innovation — helping researchers identify connections across fields that might traditionally be treated as separate. Traditional approaches to interdisciplinary work often involve collaboration between specialists from different fields, each bringing their domain-specific expertise to a common problem. While this approach can yield valuable insights, it often maintains the conceptual boundaries between disciplines, missing deeper connections that might transcend traditional categories.

Universal pattern approaches address this limitation through several innovations:

- **Pattern-based knowledge organization:** rather than organizing knowledge by traditional disciplinary boundaries, universal pattern approaches organize understanding around patterns that appear across different domains, creating new conceptual frameworks that transcend conventional categories.

- **Cross-domain insight generation:** by identifying similarities between patterns in different fields, universal pattern approaches can generate insights that wouldn't be visible through domain-specific analysis alone — connections that might reveal deeper principles underlying seemingly diverse phenomena.
- **Boundary phenomenon identification:** universal pattern approaches can identify phenomena that emerge at the boundaries between traditional disciplines — processes that might be missed when each field is analyzed in isolation.
- **Integrated problem framing:** rather than approaching problems through the lens of specific disciplines, universal pattern approaches can frame challenges in terms of patterns that span across different domains, enabling more integrated and comprehensive solutions.

During my sabbatical in São Paulo, I had the opportunity to participate in interdisciplinary research initiatives that were exploring these pattern-based approaches. What struck me was how the recognition of similar patterns across different fields could transform not just specific research questions but the entire conceptual framework through which problems were understood and addressed. One particularly captivating application was in sustainability research, where universal pattern approaches helped identify similar organizational principles across ecological, economic, and social systems.

By recognizing how similar patterns of resource distribution, feedback regulation, and adaptive response appeared across these different domains, researchers could develop more integrated approaches to sustainability challenges that respected the connections between environmental, economic, and social dimensions. The universal pattern approach suggests that effective interdisciplinary work isn't just about bringing together specialists from different fields but about recognizing the patterns that connect across different domains of knowledge — patterns that might reveal deeper principles underlying seemingly diverse phenomena.

Enhanced Creativity and Problem-Solving

Another promising application area is enhanced creativity and problem-solving — helping individuals and teams generate novel insights and solutions by recognizing and applying patterns across different domains. Traditional approaches to creativity often rely on domain-specific expertise or general brainstorming techniques. While these approaches can be valuable, they often miss the potential for creative insights that come from recognizing how patterns in one domain might be applied to challenges in completely different contexts. Universal pattern approaches offer several advantages for creativity and problem-solving:

- **Cross-domain pattern transfer:** by recognizing similarities between patterns in different domains, universal pattern approaches can facilitate the transfer of solutions from one context to another — applying principles that have proven effective in one field to challenges in completely different areas.

- **Abstract pattern application:** by extracting abstract templates that capture the essential structure of patterns across different manifestations, universal pattern approaches can generate more flexible and adaptable solutions that aren't limited to specific contexts.
- **Pattern-based problem reframing:** universal pattern approaches can help reframe problems in terms of more general patterns, enabling access to a wider range of potential solutions that might not be visible when challenges are framed in domain-specific terms.
- **Creative recombination across domains:** by recognizing patterns across different contexts, universal pattern approaches can generate novel combinations that apply structures from one domain to others in ways that create innovative solutions.

During my work with AI systems, I became interested in how these universal pattern approaches might enhance both human and artificial creativity. The key insight was that creative breakthroughs often come not from generating completely new patterns but from recognizing how existing patterns in one domain might be applied to challenges in another. This perspective suggests that enhanced creativity isn't just about generating more ideas but about developing better pattern recognition across different domains — about seeing the connections between seemingly unrelated fields that might reveal new approaches to persistent challenges.

Integrated Education and Learning

A third promising application area is integrated education and learning — helping students develop more coherent understanding across traditionally separate subjects and disciplines. Traditional educational approaches typically organize learning into separate subjects like mathematics, science, language, history, and art, each with its own curriculum, methods, and standards. While this subject-based organization can provide depth in specific areas, it often creates artificial boundaries between fields that are deeply interconnected in both theory and practice. Universal pattern approaches offer several advantages for education and learning:

- **Pattern-based knowledge integration:** rather than treating subjects as separate domains, universal pattern approaches organize learning around patterns that appear across different fields, helping students recognize connections between seemingly unrelated areas.
- **Transfer learning across subjects:** by identifying similarities between patterns in different subjects, universal pattern approaches can help students transfer understanding from one area to another, accelerating learning across different domains.
- **Boundary concept exploration:** universal pattern approaches can help students explore concepts that emerge at the boundaries between traditional subjects — ideas that might be missed when each field is taught in isolation.
- **Integrated problem-based learning:** rather than approaching problems through the lens of specific subjects, universal pattern approaches can frame

challenges in terms of patterns that span across different domains, enabling more integrated and comprehensive learning experiences.

During my early retirement and sabbatical, I've had the opportunity to explore how these universal pattern approaches might transform education at various levels. What struck me was how the recognition of similar patterns across different subjects could help students develop not just specific knowledge but a more coherent and integrated understanding of the world. One particularly interesting application was in STEAM education (Science, Technology, Engineering, Arts, and Mathematics), where universal pattern approaches helped students recognize connections between scientific principles, technological applications, engineering designs, artistic expressions, and mathematical structures.

By identifying how similar patterns appeared across these different domains, students could develop a more integrated understanding that transcended traditional subject boundaries. The universal pattern approach suggests that effective education isn't just about mastering separate subjects but about recognizing the patterns that connect across different domains of knowledge — patterns that might help students develop a more coherent and integrated understanding of the world.

Comparison with Traditional Approaches

Across these applications, several patterns emerge when comparing universal pattern approaches with traditional methods of research, creativity, and education:

- **Integration quality:** universal pattern approaches typically achieve more coherent integration across different domains, particularly for complex challenges that span traditional disciplinary boundaries.
- **Transfer effectiveness:** approaches based on universal patterns typically enable more effective transfer of insights and solutions across different contexts, leveraging similarities to accelerate learning and innovation across domains.
- How might your own learning process reflect these fractal principles? Think about how you integrate new information across different levels of understanding.
- **Boundary phenomenon identification:** universal pattern approaches are typically more effective at identifying phenomena that emerge at the boundaries between traditional categories, revealing insights that might be missed when each domain is analyzed in isolation.
- **Creative recombination value:** approaches based on universal patterns typically generate more valuable novel combinations by applying structures from one domain to others in ways that create innovative solutions to persistent challenges.

These comparative advantages aren't marginal but often substantial, particularly for complex challenges that span across different domains and require integrated understanding rather than fragmented expertise. During my three decades thinking about these issues and experimenting with different approaches to knowledge and innovation, I've become convinced that the universal pattern approach represents not just an incremental improvement but a fundamental reconceptualization of how we understand and address complex challenges — one that better aligns with the integrated nature of reality itself.

Why It Matters

Transforming Our Understanding

The universal pattern approach isn't just theoretically fascinating but has profound implications for how we understand and engage with the world around us.

Enhancing Integrative Consciousness

Traditional approaches to knowledge often focus on specific domains without developing broader awareness of the patterns that connect across different fields. This fragmented approach can lead to specialized expertise in isolated areas without transforming our overall understanding of the world. The universal pattern perspective suggests a different path to intellectual development — one centered on enhancing integrative consciousness:

- **Pattern awareness across domains:** developing the ability to recognize patterns that span across different fields and contexts, seeing how similar structures and principles might appear in seemingly unrelated areas.
- **Connection recognition:** learning to identify the relationships between patterns in different domains, recognizing how processes in one area might influence or constrain possibilities in others.
- **Boundary awareness:** developing sensitivity to phenomena that emerge at the boundaries between traditional categories, recognizing insights that might be missed when each domain is analyzed in isolation.

- **Integrated understanding:** cultivating the capacity to maintain coherent understanding across different domains rather than fragmenting knowledge into separate compartments based on conventional disciplinary boundaries.

During my sabbatical in São Paulo, I explored how these aspects of integrative consciousness might be cultivated through specific practices and frameworks. The key insight was that developing this consciousness isn't just about acquiring knowledge in multiple fields but about training perception — learning to see patterns and connections that were previously invisible or only partially recognized. This enhanced integrative consciousness doesn't just improve specific insights or solutions but transforms our entire relationship with knowledge — helping us see the world not as a collection of separate domains but as an integrated pattern that manifests differently across diverse contexts.

Implications for Knowledge Organization

The universal pattern perspective also has profound implications for how we organize and structure knowledge — suggesting possibilities that go beyond traditional disciplinary categories and hierarchical classification systems. Several alternative organization principles emerge from the universal pattern approach:

- **Pattern-based organization:** knowledge might be organized around patterns that appear across different domains rather than being fragmented into separate fields based on specific content or context.

- **Network structures:** rather than hierarchical classification systems, knowledge might be organized through network structures that represent the connections between patterns across different domains and scales.
- **Dynamic organization:** knowledge organization might be dynamic rather than static, evolving in response to new insights about the patterns that connect across different domains and contexts.
- **Integrated frameworks:** rather than maintaining sharp boundaries between different fields, knowledge might be organized through integrated frameworks that recognize the connections between patterns across traditionally separate domains.

During my work with AI systems since founding zyrian.ai in 2013, I became interested in how these alternative organization principles might transform approaches to knowledge representation and retrieval. Traditional knowledge systems often maintain the conventional boundaries between disciplines, missing the potential for more integrated understanding based on the patterns that connect across different domains. The universal pattern perspective suggests that effective knowledge organization wouldn't just categorize information into separate fields but would represent the connections between patterns across different domains — connections that might reveal deeper principles underlying seemingly diverse phenomena.

Ethical Considerations in Pattern Recognition

The potential of universal pattern approaches brings extraordinary ethical responsibilities, particularly as these approaches are implemented in technologies that influence how we understand and engage with the world. Several principles should guide the development of pattern recognition technologies:

- **Diversity preservation:** technologies should enhance rather than reduce the diversity of patterns across different domains, recognizing the value of both similarities and differences across contexts.
- **Appropriate abstraction:** development should respect the balance between abstraction and specificity, avoiding both overgeneralization that misses crucial differences and excessive fragmentation that misses meaningful connections.
- **Transparent pattern identification:** the basis for identifying patterns across domains should be transparent and understandable, allowing for evaluation and potential adjustment of how similarities are recognized and applied.
- **Cultural context respect:** technologies should respect how patterns are recognized and interpreted differently across cultural contexts, avoiding the imposition of pattern frameworks from one tradition onto others.
- **Balanced integration:** technologies should support balanced integration across different domains rather than imposing rigid unification that might miss the distinctive value of different fields and perspectives.

During my early retirement and sabbatical, I've had time to reflect on how these ethical considerations apply to technologies that recognize and apply patterns across different domains. The most effective technologies are those that enhance rather than reduce the richness of human understanding — that help us recognize meaningful connections while respecting the distinctive value of different domains and perspectives. The universal pattern perspective suggests that these ethical considerations aren't separate from technical development but integral to it — that the patterns of diversity preservation, appropriate abstraction, transparency, cultural respect, and balanced integration we design into our systems will shape not just their performance but their impact on human understanding and well-being.

Practical Framework

"Universal Pattern" Technique

To help you apply these concepts in your own exploration of patterns across domains, I offer this practical framework for universal pattern recognition inspired by the principles discussed in this chapter.

Step 1

Recognize Patterns Across Domains

The foundation of universal pattern recognition is identifying similarities across different fields and contexts. Practice recognizing patterns across domains that are traditionally treated as separate:

- **Structural patterns:** notice recurring organizational structures across different domains. How do network patterns appear in neural systems, language, social relationships, and physical processes? How do hierarchical structures manifest across different fields? What common principles might underlie these similar forms?
- **Functional patterns:** observe recurring functional principles across different contexts. How do similar approaches to information processing, resource distribution, or adaptive response appear in different systems? What common challenges might these functional similarities address?
- **Temporal patterns:** identify recurring temporal structures across different domains. How do similar patterns of development, cycle, or transformation appear in different contexts? What common principles might govern these temporal similarities?
- **Informational patterns:** notice how information is organized and transferred across different systems. How do similar patterns of encoding, processing, and communication appear in different domains? What common principles might underlie these informational similarities?

During my visits to Niagara Falls and Foz do Iguaçu, I practiced this cross-domain pattern recognition by observing how water flow patterns resembled structures in completely different domains — how turbulence patterns resembled certain neural activation patterns, how branching river systems resembled semantic networks in language, how the interaction between water and rock created boundaries with properties similar to those between different states of consciousness. This practice suggests that cross-domain patterns are often more pervasive than we realize — that with attention and practice, we can learn to recognize the similarities that connect across different aspects of reality.

Step 2

Identify Cross-Domain Connections

Once you've recognized patterns across different domains, look for connections between these patterns — how structures in one field might relate to, influence, or constrain possibilities in others:

- **Structural homologies:** how do organizational structures in different domains exhibit similar mathematical or topological properties? How might these structural similarities reflect deeper principles that transcend specific contexts?
- **Functional parallels:** how do systems in different domains solve similar functional challenges? How might these parallel solutions reflect convergence toward optimal approaches for specific functions?
- **Temporal isomorphisms:** how do processes in different domains exhibit similar patterns of change over time? How might these temporal similarities reflect deeper principles of development or transformation?
- **Information flow similarities:** how does information move through systems in different domains? How might these similar patterns of information transfer reflect deeper principles of communication or coordination?

During my early explorations with pattern recognition, I became fascinated by how similar network structures appeared in contexts as diverse as neural systems, language networks, social relationships, and physical processes. These explorations suggested that certain network properties might represent optimal solutions to challenges that appear across different domains — challenges like balancing efficiency and robustness, local specialization and global integration, stability and adaptability. This practice suggests that cross-domain connections aren't just theoretical constructs but observable aspects of reality — that with attention and appropriate frameworks, we can identify and characterize how patterns in different domains relate to each other through various forms of structural, functional, temporal, and informational similarity.

Step 3

Extract Abstract Templates

The next step is extracting abstract templates that capture the essential structure of patterns across different manifestations — representations that maintain core properties while allowing for variation in specific details:

- **Identify essential features:** what aspects of patterns remain consistent across different manifestations? What properties define the core structure regardless of the specific domain or context?
- **Allow for flexible variation:** what aspects of patterns vary across different manifestations? What properties adapt to the specific conditions of each domain or context?

- **Balance abstraction levels:** what level of abstraction best captures the pattern across different domains? How can the template maintain enough specificity to be meaningful while allowing enough flexibility to apply across diverse contexts?
- **Represent relationships between features:** how do different aspects of the pattern relate to each other? What relationships between features remain consistent across different manifestations?

During my work with pattern recognition systems, I've found that effective abstract templates aren't rigid structures but flexible frameworks that capture essential relationships while allowing for adaptation to specific contexts. The most valuable templates maintain their core structure across different domains while accommodating the particular conditions of each field. This practice suggests that abstract pattern extraction isn't just about identifying similarities but about recognizing the balance between consistency and variation — about developing representations that capture what remains the same across different manifestations while respecting how patterns adapt to the specific conditions of each domain.

Step 4

Apply Patterns Across Domains

The final step is applying patterns from one domain to others — using the abstract templates you've developed to generate insights, solutions, or approaches in new contexts:

- **Select relevant patterns:** for the specific challenge or question you're addressing, identify which patterns from other domains might be most relevant. Which abstract templates might offer valuable perspectives or approaches?

- **Adapt to target domain:** how might the selected patterns need to be adapted to the specific conditions of the target domain? What aspects would need to be modified to fit the particular context?
- **Generate novel combinations:** how might patterns from different domains be combined to create new approaches? What innovative solutions might emerge from integrating structures from diverse fields?
- **Evaluate applications:** how effectively do the applied patterns address the challenges or questions in the target domain? What insights or solutions do they generate that might not have been visible through domain-specific approaches alone?

During my three decades thinking about these issues, I've observed how this cross-domain pattern application has transformed approaches to various challenges — not through imposing abstract frameworks but through recognizing and adapting patterns that have proven effective in other contexts. The key insight is that many challenges we face aren't unique to specific domains but represent different manifestations of similar underlying patterns — that solutions developed in one field might be adapted to address seemingly unrelated challenges in completely different contexts.

Call to Action

Exploring Your Universal Patterns

As we conclude this chapter on universal patterns, I invite you to begin applying these insights to enhance your own understanding of the connections across different domains. Here are specific actions you can take:

- Practice cross-domain observation. Set aside time regularly to notice patterns that appear across different fields and contexts. Keep a pattern journal where you record these observations, looking for similarities that connect across different aspects of reality.
- Experiment with abstract template creation. When you recognize similar patterns across different domains, try to extract the essential structure that remains consistent across these diverse manifestations. Practice developing flexible templates that capture core properties while allowing for contextual variation.
- Develop cross-domain awareness in your learning and problem-solving. When encountering new information or challenges, practice considering how patterns from completely different domains might offer valuable perspectives or approaches. Look for insights that might not be visible through domain-specific frameworks alone.
- Initiate pattern conversations with friends or colleagues from different fields. Share the concept of universal patterns and explore together what connections you've noticed across different domains. These conversations can reveal insights that might not be visible through individual observation alone.

The universal pattern approach isn't just a theoretical framework but a practical method for enhancing understanding and generating innovative solutions. By recognizing and leveraging the patterns that connect across different domains, you can develop a more integrated perspective that transcends traditional boundaries between fields and contexts. In the next chapter, we'll explore the ethical implications and future directions of Fractal Neural Architectures — how the principles we've examined throughout this book might shape the development of technology, society, and human understanding in the years and decades ahead.

CHAPTER 10

Ethical Considerations and Future Implications

Narrative Opening: The Resonance of Responsibility

The monastery courtyard was silent except for the deep, resonant tone of the singing bowl. As Kai ran the wooden mallet around its rim, the sound grew in intensity, filling the space with vibrations that seemed to penetrate beyond the physical. Elena closed her eyes, feeling the sound waves wash over her. "It's remarkable how something so simple can create such complex patterns", she whispered when the resonance finally faded. "The vibrations create standing waves — fractal patterns in the air itself." Professor Chen nodded thoughtfully. "What's even more remarkable is how these patterns connect across different domains — physical vibrations, neural responses, conscious experience, cultural traditions.

The same essential structure manifests in different forms depending on context." An elderly monk who had been observing them spoke softly. "In our tradition, we see these patterns as more than physical phenomena. They represent the interconnection of all things — the way seemingly separate domains are united through common principles that transcend conventional boundaries." "That's precisely what we've been exploring through our work on Fractal Neural Architectures",

Kai replied. "How similar patterns appear across neural networks, language, consciousness, relationships, physical laws — suggesting deeper connections that traditional disciplinary boundaries might miss." The monk smiled.

"Different paths, similar destinations. Your scientific approach and our contemplative tradition are exploring the same fundamental reality from different perspectives." "But with these insights comes responsibility", Elena added. "If everything is connected through these fractal patterns, then our actions ripple across domains in ways we might not immediately recognize. Understanding these patterns isn't just an intellectual exercise but an ethical imperative." Professor Chen nodded. "That's the challenge we face. As we develop technologies based on these fractal principles — systems that can recognize and leverage patterns across different domains — we need to consider not just what they can do but what they should do. Not just their capabilities but their implications." "And who decides?" asked a younger monk who had joined them.

"If these technologies transform how we understand and engage with the world, who determines how they're developed and applied? Who ensures they serve human flourishing rather than undermining it?" "That's the question at the heart of our work", Kai acknowledged. "We're not just developing new technologies but new frameworks for thinking about their ethical implications — approaches that recognize how actions in one domain ripple across others in ways that traditional ethical frameworks might miss." The elderly monk lifted the singing bowl. "Perhaps we might begin with a moment of reflection — a silent prayer for wisdom in navigating these profound questions." As the bowl's resonance filled the courtyard again, they each contemplated the patterns that connected across different domains of knowledge and experience — and the responsibility that came with recognizing these connections.

The vibrations weren't just physical phenomena but metaphors for how insights and actions ripple across domains, creating patterns of influence that extend far beyond their immediate context. In that moment of shared contemplation, scientific innovation and spiritual tradition weren't separate domains but different perspectives on the same fundamental reality — a reality where everything is connected through patterns that transcend conventional boundaries, where understanding these connections brings not just knowledge but responsibility.

Have you noticed similar structures appearing in seemingly unrelated fields? This question lies at the heart of our exploration of fractal neural architectures.

Ethical Dimensions of Fractal Thinking

The narrative above introduces a profound reality: recognizing the fractal patterns that connect across different domains isn't just an intellectual exercise but an ethical imperative. It transforms not just how we understand the world but how we engage with it — raising questions about responsibility, values, and human flourishing that traditional ethical frameworks might not fully address.

The Ethics of Interconnection

Traditional ethical frameworks often focus on specific domains — bioethics for medical questions, environmental ethics for ecological concerns, information ethics for technological issues — treating each area as a separate sphere with its own principles and considerations. This domain-specific approach misses how actions in one area ripple across others, creating patterns of influence that transcend conventional boundaries. The fractal perspective suggests a different approach through the ethics of interconnection — recognizing how actions in one domain influence others through patterns that connect across different fields and scales. Several key principles emerge from this interconnected approach:

- **Cross-domain responsibility:** ethical consideration must extend beyond immediate effects in specific domains to include how actions ripple across different fields and scales. Decisions about technology development, for instance, must consider not just technical performance but implications for consciousness, relationships, social systems, and physical environments.
- **Pattern awareness:** ethical reasoning must include awareness of the patterns that connect across different domains — recognizing how similar structures might appear in contexts that traditional frameworks treat as separate, revealing connections that might be missed when each area is analyzed in isolation.
- **Scale integration:** ethical consideration must integrate across different scales — from immediate effects to long-term implications, from individual impacts to collective consequences, from local influences to global patterns — recognizing how actions at one scale ripple across others in ways that might not be immediately apparent.
- **Boundary transcendence:** ethical frameworks must transcend traditional boundaries between disciplines and domains — recognizing how actions in one area

> influence others through connections that conventional categories might obscure, creating responsibilities that extend beyond traditional jurisdictions.

During my visit to a monastery in China, I had a profound experience that illuminated these principles of interconnection. While my business partner received acupuncture treatment, I was invited to play a giant singing bowl. As I struck the bowl and it resonated beautifully, I made a silent prayer about the wars happening around the world.

In that moment, I experienced a brief loss of consciousness, and when I returned to awareness, I saw a vision of a monkey waving a small smoking bag with a big smile, saying that everything is fine. This experience, strange as it might seem, highlighted the connections across different domains — physical vibrations, altered consciousness, symbolic imagery, and global concerns about violence and conflict. It suggested that understanding the patterns of violence in human societies might require perspectives that transcend traditional boundaries between physical, psychological, social, and even spiritual domains.

The ethics of interconnection suggests that addressing complex challenges like war and violence requires approaches that recognize how these patterns connect across different domains — how physical, psychological, social, economic, and political factors interact through feedback loops that transcend conventional categories, creating patterns that can't be fully understood or addressed through domain-specific approaches alone.

Responsibility in Pattern Recognition

As we develop technologies that can recognize and leverage patterns across different domains, we face profound questions about responsibility — about who determines how these capabilities are developed and applied, who ensures they serve human flourishing rather than undermining it. Several key considerations emerge around responsibility in pattern recognition:

- **Representation in development:** who participates in developing pattern recognition technologies? Whose perspectives and values shape how patterns are identified, characterized, and applied across different domains? How can we ensure diverse participation that reflects the full range of human experience and wisdom?
- **Transparency in operation:** how can pattern recognition systems be made transparent and understandable? How can we ensure that the basis for identifying patterns across domains is open to evaluation and potential adjustment rather than operating as a black box that evades accountability?
- **Accountability for impacts:** who's responsible for the effects of pattern recognition technologies across different domains? How can we ensure accountability for both intended and unintended consequences as these technologies influence how we understand and engage with the world?
- **Governance across boundaries:** how should pattern recognition technologies be governed when their impacts transcend traditional jurisdictional boundaries? What frameworks might enable responsible development and application across different domains and contexts?

During my work with AI systems since founding zyrian.ai in 2013, I've become increasingly aware of how these questions of responsibility apply to technologies that recognize and leverage patterns across different domains.

The most challenging aspects aren't technical but ethical — determining not just what these systems can do but what they should do, not just their capabilities but their implications. The fractal perspective suggests that responsibility in pattern recognition can't be confined to any single domain but must extend across the different fields and scales that these technologies influence. It requires frameworks that can integrate across traditionally separate areas like technical development, psychological impact, social consequences, and environmental implications — recognizing how actions in one domain ripple across others in ways that conventional approaches might miss. This integrated approach to responsibility doesn't eliminate the need for domain-specific expertise but complements it with recognition of the patterns that connect across different fields — patterns that might reveal ethical considerations that would be missed when each domain is analyzed in isolation.

Values Across Domains

A third ethical dimension of fractal thinking concerns values — the principles and priorities that guide how pattern recognition technologies are developed and applied across different domains. Traditional approaches to values often focus on specific domains — efficiency and performance in technical systems, autonomy and beneficence in healthcare, justice and fairness in social systems — treating each area as a separate sphere with its own normative frameworks. This domain-specific approach misses how values interact across different fields and scales, creating patterns of influence that transcend conventional boundaries.

Several key considerations emerge around values across domains:

- **Value integration:** how can we develop value frameworks that integrate across different domains rather than treating each area as a separate sphere with its own normative principles? How might values like justice, compassion, sustainability, and wisdom manifest across different fields and scales?
- **Value conflicts across domains:** how should we address situations where values in one domain conflict with those in another? How can we develop approaches that recognize these tensions without reducing all values to a single metric or framework?
- **Cultural diversity in values:** how can we respect cultural diversity in values while recognizing patterns that might transcend specific traditions? How might different cultural perspectives contribute to more integrated understanding of values across domains?
- **Evolution of values:** how might values evolve as we recognize new patterns across different domains? How can value frameworks remain open to revision and refinement as our understanding of cross-domain connections develops?

During my visits to China, particularly in remote regions near the Kazakhstan frontier, I had fascinating discussions with scholars and practitioners from different traditions about how values manifest across different domains. What struck me was how certain core values appeared across diverse contexts — how principles like harmony, balance, and interconnection manifested in fields from medicine to ecology to social organization to spiritual practice. These cross-cultural experiences suggested that while specific expressions of values might vary across different traditions, certain core principles might transcend particular cultural contexts — reflecting deeper patterns in how human communities organize themselves across different domains and scales.

The fractal perspective suggests that effective value frameworks wouldn't just apply different principles to separate domains but would recognize how similar values might manifest across diverse contexts — how principles like justice, compassion, sustainability, and wisdom might appear in different forms across technical, psychological, social, and environmental domains while maintaining their essential structure.

Human Flourishing in a Fractal World

Perhaps the most fundamental ethical dimension of fractal thinking concerns human flourishing — how pattern recognition across domains might enhance or diminish the conditions for meaningful and fulfilling human lives. Traditional approaches to human flourishing often focus on specific domains — physical health, psychological well-being, social connection, economic prosperity — treating each area as a separate sphere with its own metrics and interventions. This domain-specific approach misses how flourishing emerges from the interaction between processes across different fields and scales, creating patterns that transcend conventional boundaries. Several key considerations emerge around human flourishing in a fractal world:

- **Integrated well-being:** how might human flourishing be understood as an integrated phenomenon that spans across physical, psychological, social, and environmental domains rather than as separate forms of well-being in different areas? How do patterns in one domain influence flourishing in others?
- **Flourishing across scales:** how does human flourishing manifest across different scales — from momentary experiences to lifetime development, from individual well-being to collective flourishing, from local communities to global systems? How do patterns at one scale influence flourishing at others?
- **Diversity in flourishing:** how can we respect diverse conceptions of flourishing across different cultural and individual contexts while recognizing patterns that might transcend specific traditions? How might different perspectives contribute to more integrated understanding of human flourishing across domains?
- **Evolution of flourishing:** how might conceptions of human flourishing evolve as we recognize new patterns across different domains? How can frameworks for understanding flourishing remain open to revision and refinement as our understanding of cross-domain connections develops?

During my early retirement and sabbatical, I've had time to reflect on how these considerations apply to my own understanding of flourishing.

What gradually became clear was how my own well-being emerged from the interaction between processes across different domains — how physical health influenced psychological experience, how social connection shaped environmental engagement, how spiritual meaning informed economic choices, creating patterns of flourishing that transcended conventional boundaries between separate domains. The fractal perspective suggests that human flourishing isn't just the sum of separate forms of well-being in different domains but emerges from the patterns that connect across them — from the integration of physical, psychological, social, environmental, and spiritual processes through feedback loops that transcend conventional boundaries between separate areas of human experience and engagement.

Contrarian Perspective

Beyond Traditional Ethical Frameworks

The fractal approach to ethics challenges several fundamental assumptions in traditional ethical frameworks. Let me explain why I believe some established approaches miss the connections that integrate across different domains of ethical consideration.

Why Domain-Specific Ethics Falls Short

Traditional approaches to ethics often rely on domain-specific frameworks — bioethics for medical questions, environmental ethics for ecological concerns, information ethics for technological issues — each with its own principles, methods, and standards. While this approach has yielded valuable insights within specific domains, it often misses the patterns that connect across different fields. Several limitations of domain-specific ethics have become increasingly apparent:

- **Boundary blindness:** specialized ethical frameworks often develop blind spots at their boundaries, missing considerations that emerge from the interaction between processes typically addressed by different fields.
- **Conceptual isolation:** domain-specific approaches typically develop their own specialized concepts and terminology, creating artificial barriers to recognizing similar ethical patterns that might be described through different language in different fields.
- **Methodological constraints:** each ethical domain typically develops its own methodological approaches, limiting the ability to recognize patterns that might require integrated methods spanning across traditional boundaries.
- **Fragmented understanding:** perhaps most fundamentally, domain-specific ethics fragments our understanding of ethical responsibility into separate spheres, missing the patterns that integrate across different aspects of human action and consequence.

During my travels and cross-cultural experiences, I've observed how different ethical traditions approach these limitations in different ways. Western ethics has traditionally emphasized domain-specific frameworks, while many Eastern and indigenous traditions have maintained more integrated approaches that recognize connections across domains that Western frameworks treat as separate. The fractal alternative doesn't eliminate the value of domain-specific insights but complements them with recognition of the patterns that connect across different fields — patterns that might be missed when each domain is treated as a separate ethical universe.

It suggests that the most complete ethical understanding comes not from reducing all considerations to a single framework but from recognizing how similar patterns manifest differently across diverse contexts. This perspective aligns with emerging transdisciplinary approaches to ethics that seek to transcend traditional boundaries between fields. It suggests that ethical consideration might be better organized around patterns that appear across different domains rather than being fragmented into separate frameworks based on the specific content being addressed.

The Myth of Ethical Autonomy

Another limitation of traditional approaches is the assumption of ethical autonomy — the idea that different ethical domains operate independently from each other, with clear boundaries between areas like medical ethics, environmental ethics, and information ethics. This assumption has led to the fragmentation of ethical consideration into separate frameworks, each with its own principles, methods, and standards. Several problems with the ethical autonomy assumption have become apparent:

- **Interconnected consequences:** the consequences of actions aren't confined to any single domain but ripple across different fields and scales — technical decisions influence psychological experiences, psychological impacts shape social dynamics, social patterns affect environmental systems, creating feedback loops that transcend conventional ethical categories.
- **Boundary phenomena:** many of the most crucial ethical considerations emerge at the boundaries between

traditional domains — phenomena that can't be fully addressed through the principles and methods of any single ethical framework.

- **Common patterns across domains:** similar ethical patterns often appear across domains that are traditionally treated as separate, suggesting deeper connections that transcend conventional boundaries between fields like bioethics, environmental ethics, and information ethics.
- **Integrated human experience:** human experience integrates across domains that ethical frameworks treat as separate, creating a gap between lived reality and fragmented ethical consideration.

During my work with AI systems, I became interested in how these cross-domain connections might inform more effective approaches to technological ethics. Traditional ethical frameworks often focus on specific domains or considerations, missing the integrated nature of ethical responsibility that spans across areas like technical performance, psychological impact, social consequences, and environmental implications. The fractal perspective suggests that effective ethical frameworks wouldn't just apply different principles to separate domains but would recognize how similar considerations might manifest across diverse contexts — how principles like justice, compassion, sustainability, and wisdom might appear in different forms across technical, psychological, social, and environmental domains while maintaining their essential structure.

Challenging the Principle-Application Dichotomy

Perhaps the most fundamental limitation of traditional approaches is the sharp dichotomy between ethical principles and practical application — the idea that abstract ethical frameworks exist separately from concrete ethical decisions, with clear boundaries between theoretical ethics and practical moral reasoning. This dichotomy has led to the separation of ethical theory from practical wisdom, creating artificial barriers between those who develop abstract frameworks and those who apply ethical reasoning in specific contexts. Several problems with the principle-application dichotomy have become apparent:

- **Embodied ethical knowledge:** much of our most valuable ethical understanding isn't abstract principle but embodied wisdom that emerges from practical engagement with moral challenges — knowledge that can't be fully captured through theoretical frameworks alone.
- **Principle-practice feedback:** ethical principles both inform and are informed by practical application, creating feedback loops that transcend the conventional boundary between theory and practice.
- **Pattern recognition across contexts:** some of the most valuable ethical insights come from recognizing patterns across different contexts of moral decision-making — patterns that might not be visible through theoretical analysis alone.
- **Integrated knowing-doing:** the most effective ethical approaches often integrate knowing and doing rather

than treating them as separate domains — approaches that recognize how ethical understanding emerges through active engagement rather than passive reflection.

During my sabbatical in São Paulo, I had fascinating discussions with practitioners across different fields about how they integrated ethical principles with practical wisdom. What struck me was how the most effective approaches didn't treat theory and practice as separate domains but recognized the patterns that connected across different forms of ethical knowing and doing. The fractal perspective suggests that ethical understanding isn't just abstract principle but emerges from the recognition of patterns across different contexts of moral engagement.

It proposes that the most complete ethical understanding comes not from separating principles from application but from recognizing how similar patterns manifest across both theoretical frameworks and practical decisions. This approach doesn't eliminate the distinction between different forms of ethical knowledge but recognizes how they might be connected through common patterns that transcend traditional boundaries between theory and practice, abstract and concrete, knowing and doing.

Beyond the Individual-Collective Dichotomy

A fourth limitation of traditional approaches is the sharp dichotomy between individual and collective ethics — the idea that there's a clear boundary between personal moral responsibility and collective ethical consideration, with different principles and frameworks applying to each domain. This dichotomy has led to the separation of individual ethics from collective responsibility, creating artificial barriers between personal moral reasoning and social ethical frameworks. Several problems with the individual-collective dichotomy have become apparent:

- **Scale-transcending patterns:** ethical patterns often transcend the conventional boundary between individual and collective — similar principles might apply across different scales from personal decisions to global systems, though with important variations that reflect the specific conditions of each context.
- **Individual-collective feedback:** individual actions both influence and are influenced by collective patterns, creating feedback loops that transcend the conventional boundary between personal and social ethics.
- **Emergent ethical properties:** some of the most crucial ethical considerations emerge from the interaction between individual actions and collective patterns — properties that can't be fully captured through either individual or collective frameworks alone.
- **Integrated ethical identity:** human ethical identity integrates across individual and collective dimensions, creating a gap between lived moral experience and frameworks that maintain a sharp boundary between personal and social ethics.

During my early retirement and sabbatical, I've had time to explore these issues through both theoretical reflection and practical observation. What gradually became clear was how ethical patterns might transcend the individual-collective dichotomy — how similar principles might apply across different scales from personal decisions to global systems, though with essential variations that reflect the specific conditions of each context. The fractal perspective suggests that ethical patterns might be scale-invariant — maintaining their essential structure while adapting to different contexts across the spectrum from individual to collective. It proposes that effective ethical frameworks wouldn't just apply different principles to personal and social domains but would recognize how similar patterns might manifest across these different scales.

This approach doesn't eliminate the practical distinction between individual and collective ethics but recognizes how they might be connected through common patterns that transcend traditional boundaries between personal and social, private and public, individual and collective.

FNA Approach to Ethical Pattern Recognition

The fractal perspective on ethics isn't just a theoretical framework but the foundation for practical approaches using Fractal Neural Architectures. Let's explore how FNAs might address aspects of ethical pattern recognition and integration across domains.

Cross-Domain Ethical Pattern Recognition

Traditional approaches to ethical reasoning typically focus on specific domains — either bioethical considerations, environmental concerns, or technological implications — using different frameworks and principles for each domain. This domain-specific focus misses how similar ethical patterns might appear across different contexts and how recognition in one domain might inform understanding in others. FNAs take a fundamentally different approach through cross-domain ethical pattern recognition — identifying and characterizing ethical patterns that appear across different domains and contexts. The key insight is that ethical reasoning can't be confined to any single domain but emerges from the ability to identify similar considerations across different contexts — to recognize how patterns in one domain might have parallels in others, revealing deeper principles that transcend specific content.

In an FNA approach to cross-domain ethical pattern recognition: ethical patterns are represented through abstract templates that capture their essential structure while allowing for variation in specific manifestations across different domains. Do you see how these patterns repeat across different domains? This question lies at the heart of our exploration of fractal neural architectures. Recognition occurs simultaneously across multiple domains, with ethical patterns in one context informing and constraining recognition in others through bidirectional information flow. The system identifies not just structural similarities but functional parallels — similar ethical principles that address similar challenges across different domains. Pattern templates evolve through exposure to diverse manifestations across different contexts, becoming more abstract and flexible while maintaining their essential structure.

This cross-domain recognition enables several capabilities that are daunting to achieve with traditional domain-specific approaches:

- **Transfer learning across ethical domains:** the system can apply ethical patterns recognized in one domain to facilitate reasoning in others, leveraging similarities to accelerate ethical learning across different contexts.
- **Abstract ethical pattern extraction:** by identifying similarities across diverse manifestations, the system can extract more abstract ethical templates that capture essential considerations while allowing for variation in specific expressions.
- **Anomaly detection through cross-domain comparison:** the system can identify ethical anomalies by comparing patterns across different domains, recognizing when a consideration in one context deviates from similar patterns observed in others.
- **Creative ethical recombination across domains:** by recognizing patterns across different contexts, the system can generate creative recombinations that apply ethical frameworks from one domain to others in novel but meaningful ways.

During my work with AI systems since founding zyrian.ai in 2013, I've been particularly interested in how these cross-domain recognition capabilities might transform approaches to ethical reasoning in artificial intelligence. Traditional AI ethics often develops separate frameworks for different domains, missing the potential for transfer and integration across contexts. The fractal approach suggests that effective ethical reasoning wouldn't just recognize patterns within specific domains but would identify and leverage the patterns that appear across different contexts — patterns that might reveal deeper principles underlying seemingly diverse ethical considerations.

Integrated Multi-Domain Ethical Modeling

A particularly promising application of FNAs to ethics is integrated multi-domain ethical modeling — representing and simulating ethical systems that span across traditionally separate domains like technological development, psychological impact, social consequences, and environmental implications. Traditional ethical modeling approaches typically focus on specific domains, using different frameworks and techniques for each area. This domain-specific focus misses how real ethical challenges often integrate across domains that academic frameworks treat as separate — how technological decisions influence psychological experiences, how psychological impacts shape social dynamics, how social patterns affect environmental systems, creating feedback loops that transcend conventional boundaries.

FNAs take a fundamentally different approach through integrated multi-domain ethical modeling — representing and simulating ethical systems that span across traditionally separate domains through unified frameworks that can capture the patterns connecting different fields and scales. The key insight is that effective ethical modeling can't be confined to any single domain but must represent how considerations in different fields interact through feedback loops that transcend conventional boundaries — how decisions in one area influence outcomes in others through connections that might be missed when each domain is modeled in isolation.

In practice, implementations would include additional components like uncertainty quantification, hierarchical pattern organization, and more sophisticated similarity metrics. The specific architecture would be adapted to the particular ethical domains being integrated and the available data sources.

Evaluation Metrics for Ethical Patterns

Evaluating the quality of ethical pattern recognition requires metrics that go beyond domain-specific accuracy to assess how effectively patterns are recognized and applied across different contexts. Several specialized metrics have proven valuable:

- **Cross-domain ethical transfer accuracy:** how effectively can ethical patterns recognized in one domain be applied to facilitate reasoning in other domains? This measures the system's ability to leverage similarities across contexts.
- **Abstract ethical template quality:** how well do abstract templates capture the essential structure of ethical patterns while allowing for variation in specific manifestations? This assesses the balance between specificity and generality in ethical representations.
- **Integration coherence:** how coherently does the system integrate ethical patterns across different domains? This measures whether cross-domain connections create a coherent understanding or merely a collection of separate observations.
- **Creative ethical recombination value:** how valuable are the novel ethical combinations generated by applying patterns from one domain to others? This assesses the system's ability to generate insights through cross-domain application.
- **Boundary phenomenon capture:** how effectively does the system represent ethical considerations that emerge at the boundaries between traditional domains? This measures the ability to capture considerations that wouldn't be visible through domain-specific analysis alone.

These metrics provide a more comprehensive evaluation than domain-specific accuracy measures, helping identify specific strengths and vulnerabilities in cross-domain ethical pattern recognition. During my work with ethical reasoning systems, I've found that qualitative assessment by domain experts can also provide valuable insights, particularly when experts from different fields evaluate the same patterns from their respective perspectives. This multi-perspective evaluation can reveal aspects of ethical patterns that might not be captured by any single disciplinary framework. The fractal perspective suggests additional evaluation approaches focused on how effectively systems capture the connections across different ethical domains. Rather than treating pattern quality as a domain-specific property, we can assess how patterns facilitate understanding and prediction across traditional boundaries — how they reveal connections that might be missed when each domain is analyzed in isolation.

Future Implications and Possibilities

The fractal approach to ethics and pattern recognition opens new possibilities for how we understand and address complex challenges across different domains. Let's explore some of the most promising future directions and their implications.

Transforming Ethical Education and Development

Traditional approaches to ethical education often focus on specific domains — professional ethics for particular fields, applied ethics for specific contexts, theoretical ethics for academic study — treating each area as a separate sphere with its own principles and considerations. This domain-specific approach misses how ethical patterns connect across different fields and scales, creating artificial boundaries that fragment moral understanding. The fractal perspective suggests a different approach to ethical education — one centered on recognizing and applying patterns across different domains rather than treating each area as a separate ethical universe. Several promising directions emerge from this integrated approach:

- **Pattern-based ethical curricula:** rather than organizing ethical education around separate domains like bioethics, environmental ethics, and information ethics, curricula might be structured around patterns that appear across different fields — patterns like balancing competing values, addressing unintended consequences, or navigating ethical uncertainty.
- **Cross-domain ethical case studies:** educational materials might present case studies that span across different domains, helping students recognize how similar ethical patterns might appear in diverse contexts and how insights from one field might inform understanding in others.
- **Integrated ethical simulations:** educational technologies might create simulations that span across different domains, helping students experience how decisions in one area influence outcomes in others through feedback loops that transcend conventional ethical boundaries.
- **Pattern recognition training:** ethical education might include specific training in recognizing patterns across different domains, helping students develop the perceptual and cognitive skills needed to identify similarities across diverse contexts.

During my sabbatical in São Paulo, I had fascinating discussions with educators across different fields about how these integrated approaches might transform ethical education. The key insight was that effective moral development wouldn't just teach separate frameworks for different domains but would help students recognize and apply the patterns that connect across different fields and scales. This perspective suggests that ethical wisdom isn't just domain-specific expertise but the ability to recognize and apply patterns across different contexts — to see how insights from one field might inform understanding in others, revealing connections that might be missed when each domain is treated as a separate ethical universe.

Enhancing Cross-Cultural Ethical Understanding

Another promising direction concerns cross-cultural ethical understanding — how the recognition of patterns across different cultural traditions might enhance our ability to navigate diverse moral perspectives without either imposing rigid universalism or accepting chaotic relativism. Traditional approaches to cross-cultural ethics often oscillate between universalism (assuming that the same ethical principles apply across all cultural contexts) and relativism (assuming that ethical principles are entirely specific to particular cultural traditions). This dichotomy misses how similar ethical patterns might appear across different cultural contexts while manifesting in diverse forms that reflect specific historical and social conditions.

The fractal perspective suggests a different approach through pattern recognition across cultural traditions — identifying similarities that might reveal deeper connections while respecting the distinctive features of different ethical frameworks. Several promising directions emerge from this pattern-based approach:

- **Cross-cultural ethical pattern mapping:** research might map ethical patterns across different cultural traditions, identifying both similarities that might reflect deeper connections and differences that might reveal distinctive insights from particular perspectives.
- **Pattern-based translation across traditions:** rather than attempting direct translation between ethical concepts from different traditions, approaches might focus on identifying similar patterns that manifest in different forms across diverse cultural contexts.
- **Integrated ethical frameworks:** rather than either imposing universal principles or accepting complete relativism, approaches might develop integrated frameworks that recognize both the patterns that connect across different traditions and the distinctive features that reflect specific cultural contexts.
- **Pattern-based ethical dialogue:** cross-cultural ethical dialogue might focus on recognizing and exploring patterns that appear across different traditions rather than debating abstract principles or specific practices in isolation.

During my visits to China, particularly in remote regions near the Kazakhstan frontier, I had fascinating discussions with scholars and practitioners from different traditions about how these pattern-based approaches might transform cross-cultural ethical understanding. What struck me was how certain core patterns appeared across diverse cultural contexts — how principles like reciprocity, harmony, and respect for dignity manifested in different forms across various traditions while maintaining their essential structure.

These cross-cultural experiences suggested that while specific expressions of ethical principles might vary across different traditions, certain core patterns might transcend particular cultural contexts — reflecting deeper connections in how human communities address similar moral challenges across diverse historical and social conditions.

Addressing Complex Global Challenges

Perhaps the most promising direction concerns complex global challenges — how the recognition of patterns across different domains might enhance our ability to address issues like climate change, technological disruption, social inequality, and political polarization that span across traditionally separate fields and scales. Traditional approaches to these challenges often focus on specific domains — technological solutions for climate change, economic policies for inequality, political reforms for polarization — treating each area as a separate sphere with its own interventions and metrics. This domain-specific approach misses how these challenges emerge from the interaction between processes across different fields and scales, creating patterns that transcend conventional boundaries. The ethical dimensions of this work became personal for me when my own algorithm was used in an application I hadn't anticipated.

Seeing my research applied in contexts I hadn't considered forced me to confront the ripple effects of innovation across different domains. This experience fundamentally changed how I approach technology development, making me acutely aware of how patterns in one domain inevitably influence others in ways we might not immediately recognize. The fractal perspective suggests a different approach through integrated pattern recognition across domains — identifying how processes in different fields interact through feedback loops that span traditional boundaries, creating patterns that can't be fully understood or addressed through domain-specific approaches alone. Several promising directions emerge from this integrated approach:

- **Cross-domain challenge mapping:** research might map how complex challenges emerge from the interaction between processes across different domains — how climate change involves technological, economic, social, psychological, and political factors that interact through feedback loops spanning traditional boundaries.
- **Integrated intervention design:** rather than developing separate interventions for different domains, approaches might create integrated strategies that address patterns spanning across different fields and scales — recognizing how actions in one area influence outcomes in others through feedback loops that transcend conventional boundaries.
- **Pattern-based prediction and monitoring:** forecasting and monitoring systems might track patterns that span across different domains rather than focusing on separate metrics for each field — helping identify early warning signs that might not be visible through domain-specific analysis alone.
- **Cross-scale coordination mechanisms:** governance approaches might develop mechanisms for coordinating

actions across different scales — from local to global, from immediate to long-term — recognizing how patterns at one scale influence those at others through connections that transcend conventional boundaries.

During my early retirement and sabbatical, I've had time to reflect on how these integrated approaches might transform our ability to address complex global challenges. The key insight was that effective responses wouldn't just address separate domains independently but would recognize and leverage the patterns that connect across different fields and scales. This perspective suggests that some of the most important leverage points for addressing complex challenges might exist precisely at the boundaries between traditional domains — that effective interventions might focus on the patterns that connect across technological, economic, social, psychological, and political processes rather than treating each area as a separate sphere with its own solutions and metrics.

Comparison with Traditional Approaches

Across these future directions, several patterns emerge when comparing fractal approaches with traditional methods of ethical reasoning, cross-cultural understanding, and complex challenge response:

- **Integration quality:** fractal approaches typically achieve more coherent integration across different domains, particularly for complex challenges that span traditional disciplinary boundaries.
- **Transfer effectiveness:** approaches based on pattern recognition typically enable more effective transfer

of insights and solutions across different contexts, leveraging similarities to accelerate learning and innovation across domains.

- How might your own learning process reflect these fractal principles? Think about how you integrate new information across different levels of understanding.
- **Boundary phenomenon identification:** fractal approaches are typically more effective at identifying phenomena that emerge at the boundaries between traditional categories, revealing insights that might be missed when each domain is analyzed in isolation.
- **Creative recombination value:** approaches based on pattern recognition typically generate more valuable novel combinations by applying structures from one domain to others in ways that create innovative solutions to persistent challenges.

These comparative advantages aren't marginal but often substantial, particularly for complex challenges that span across different domains and require integrated understanding rather than fragmented expertise. During my three decades thinking about these issues and experimenting with different approaches to ethical reasoning and complex challenge response, I've become convinced that the fractal pattern approach represents not just an incremental improvement but a fundamental reconceptualization of how we understand and address complex challenges — one that better aligns with the integrated nature of reality itself.

Why It Matters

Transforming Our Approach

The fractal approach to ethics and pattern recognition isn't just theoretically fascinating but has profound implications for how we understand and address the challenges facing humanity in the coming decades.

Enhancing Ethical Wisdom

Traditional approaches to ethics often focus on specific principles or frameworks without developing broader awareness of the patterns that connect across different domains and traditions. This fragmented approach can lead to specialized expertise in isolated areas without transforming our overall ethical understanding. The fractal perspective suggests a different path to ethical development — one centered on enhancing pattern recognition across different domains and traditions:

- **Pattern awareness across ethical domains:** developing the ability to recognize ethical patterns that span across different fields and contexts, seeing how similar considerations might appear in seemingly unrelated areas.
- **Connection recognition:** learning to identify the relationships between ethical patterns in different domains, recognizing how considerations in one area might influence or constrain possibilities in others.
- How might your own learning process reflect these fractal principles? Think about how you integrate new information across different levels of understanding.
- **Boundary awareness:** developing sensitivity to ethical considerations that emerge at the boundaries between traditional categories, recognizing insights

that might be missed when each domain is analyzed in isolation.

- **Integrated ethical understanding:** cultivating the capacity to maintain coherent ethical understanding across different domains rather than fragmenting moral consideration into separate compartments based on conventional disciplinary boundaries.

During my sabbatical in São Paulo, I explored how these aspects of ethical wisdom might be cultivated through specific practices and frameworks. The key insight was that developing this wisdom isn't just about acquiring knowledge in multiple fields but about training perception — learning to see ethical patterns and connections that were previously invisible or only partially recognized. During my research, I once spent an entire weekend generating fractal patterns with different parameters. My family thought I'd gone slightly mad as I covered the walls with printouts, stepping back and forth to see the patterns at different scales. But that immersive experience gave me an intuitive feel for fractals that no textbook could provide. This enhanced ethical wisdom doesn't just improve specific insights or solutions but transforms our entire relationship with moral consideration — helping us see ethics not as a collection of separate domains but as an integrated pattern that manifests differently across diverse contexts.

Implications for Governance and Policy

The fractal perspective also has profound implications for governance and policy — suggesting possibilities that go beyond traditional approaches to regulation, coordination, and decision-making.

Several alternative governance principles emerge from the pattern recognition approach:

- **Pattern-based regulation:** rather than developing separate regulatory frameworks for different domains, governance might focus on patterns that appear across different fields — creating integrated approaches that recognize how similar considerations manifest in diverse contexts.
- **Cross-domain coordination mechanisms:** instead of separate coordination bodies for different areas, governance might develop mechanisms that span across traditional boundaries — helping align actions across domains that influence each other through feedback loops that transcend conventional categories.
- **Pattern-sensitive decision frameworks:** decision-making processes might explicitly consider how patterns connect across different domains and scales — recognizing how choices in one area influence outcomes in others through connections that might be missed when each domain is analyzed in isolation.
- **Integrated monitoring systems:** rather than separate metrics for different domains, governance might develop integrated monitoring approaches that track patterns spanning across different fields and scales —

helping identify early warning signs that might not be visible through domain-specific analysis alone.

During my work with AI systems and technology development, I became interested in how these alternative governance principles might transform approaches to complex challenges that span across traditional boundaries. The key insight was that effective governance wouldn't just regulate separate domains independently but would recognize and leverage the patterns that connect across different fields and scales. This perspective suggests that some of the most essential governance innovations might focus on the boundaries between traditional domains — developing mechanisms for coordinating actions across fields that influence each other through feedback loops that transcend conventional categories.

Human Flourishing in a Complex World

Perhaps the most fundamental implication of the fractal approach concerns human flourishing — how pattern recognition across domains might enhance our ability to create conditions for meaningful and fulfilling lives in an increasingly complex and interconnected world. Traditional approaches to human flourishing often focus on specific domains — physical health, psychological well-being, social connection, economic prosperity — treating each area as a separate sphere with its own metrics and interventions. This domain-specific approach misses how flourishing emerges from the interaction between processes across different fields and scales, creating patterns that transcend conventional boundaries.

The fractal perspective suggests a different approach through integrated pattern recognition across domains — identifying how processes in different fields interact to create conditions for meaningful and fulfilling lives, revealing connections that might be missed when each domain is analyzed in isolation.

Several promising directions emerge from this integrated approach:

- **Cross-domain flourishing maps:** research might map how human flourishing emerges from the interaction between processes across different domains — how meaningful and fulfilling lives involve physical, psychological, social, economic, environmental, and spiritual factors that interact through feedback loops spanning traditional boundaries.
- **Integrated well-being frameworks:** rather than separate metrics for different domains, approaches might develop integrated frameworks that recognize how well-being emerges from patterns spanning across different fields and scales — helping identify aspects of flourishing that might not be visible through domain-specific analysis alone.
- **Pattern-based intervention design:** instead of separate interventions for different domains, approaches might create integrated strategies that address patterns spanning across different fields and scales — recognizing how actions in one area influence outcomes in others through feedback loops that transcend conventional boundaries.
- **Cross-scale flourishing coordination:** governance approaches might develop mechanisms for coordinating actions across different scales — from individual to community to society to global — recognizing how

> patterns at one scale influence those at others through connections that transcend conventional boundaries.

During my early retirement and sabbatical, I've had time to reflect on how these integrated approaches might transform our understanding of human flourishing. The key insight was that meaningful and fulfilling lives wouldn't emerge from optimizing separate domains independently but from recognizing and leveraging the patterns that connect across different fields and scales. This perspective suggests that some of the most vital aspects of human flourishing might exist precisely at the boundaries between traditional domains — that meaningful and fulfilling lives might emerge from the patterns that connect across physical, psychological, social, economic, environmental, and spiritual processes rather than from optimizing each area as a separate sphere with its own metrics and interventions.

Practical Framework

"Ethical Pattern" Technique

To help you apply these concepts in your own ethical reasoning and decision-making, I offer this practical framework for ethical pattern recognition inspired by the principles discussed in this chapter.

Step 1

Recognize Ethical Patterns Across Domains

The foundation of ethical pattern recognition is identifying similarities across different fields and contexts. Practice recognizing ethical patterns across domains that are traditionally treated as separate:

- **Structural patterns:** notice recurring organizational structures in ethical considerations across different domains. How do similar patterns of competing values appear in technological development, personal relationships, social systems, and environmental challenges?
- **Functional patterns:** identify similar functions that ethical considerations serve across different contexts. How do principles in one domain address challenges that parallel those in other fields?
- **Contextual patterns:** recognize how similar ethical considerations manifest differently across diverse contexts. How do principles like reciprocity, harm reduction, or fairness appear in different forms across various domains?
- **Consequence patterns:** notice similarities in how actions in different domains influence outcomes across various fields and scales. How do choices in one area create ripple effects that parallel those in other contexts?

Step 2

Extract Abstract Ethical Templates

Once you've recognized patterns across different domains, practice extracting more abstract templates that capture their essential structure while allowing for variation in specific manifestations:

- **Identify core principles:** what fundamental principles underlie the specific ethical considerations you've recognized across different domains? What essential structure remains consistent despite variations in specific manifestations?
- **Characterize pattern flexibility:** how does the pattern adapt to different contexts while maintaining its essential structure? What aspects remain consistent across domains, and what features vary based on specific conditions?
- **Develop multi-level representations:** how might the pattern be represented at different levels of abstraction? What details are essential at each level, and what features can be generalized across contexts?
- **Test pattern boundaries:** where does the pattern break down or require significant modification? What are the limits of its applicability across different domains and contexts?

Step 3

Apply Ethical Patterns Across Domains

The most valuable aspect of ethical pattern recognition is applying insights from one domain to inform understanding and action in others:

- **Select source patterns:** identify ethical patterns from domains where you have deep understanding or where principles are particularly well-developed. What patterns might serve as valuable templates for application in other contexts?

- **Identify target challenges:** select ethical challenges in domains where existing approaches seem inadequate or incomplete. What problems might benefit from insights derived from patterns in other fields?
- **Map pattern elements:** carefully map elements from the source pattern to the target challenge, identifying both similarities that enable transfer and differences that require adaptation. How might the pattern be applied while respecting the specific conditions of the target context?
- **Generate multiple ethical applications:** develop multiple potential applications rather than forcing a single approach, recognizing that patterns from other domains might be applied in different ways to the same target challenge.
- **Test and refine:** test the ethical applications you've generated, refining them based on feedback and results rather than rigidly adhering to the original pattern transfer.

During my work with AI systems and technology development, I've observed how this cross-domain application can transform approaches to persistent ethical challenges — not by imposing rigid templates from one field onto another but by recognizing how similar principles might be adapted to different contexts in ways that generate novel and effective solutions. The key insight is that effective ethical pattern application isn't about forcing the same approach onto different problems but about recognizing when similar organizational principles might be appropriate across different contexts — when the fundamental challenges being addressed in different domains might share enough common features to benefit from similar approaches.

Call to Action

Exploring Ethical Patterns

As we conclude this chapter on ethical considerations and future implications, I invite you to begin applying these insights to enhance your own ethical reasoning and decision-making. Here are specific actions you can take:

- Practice cross-domain ethical observation in your daily life and work. Set aside time regularly to notice ethical patterns that appear across different fields and contexts. Keep an ethical pattern journal where you record these observations, looking for similarities that might reveal deeper connections across domains traditionally treated as separate.
- Experiment with ethical pattern transfer between different areas of knowledge or practice. Select ethical patterns you've recognized in one domain and explore how they might apply to challenges in completely different contexts. Notice both the insights and the limitations that emerge from this cross-domain application.
- Develop abstract ethical pattern awareness in your thinking and problem-solving. Practice identifying the essential principles that underlie specific ethical manifestations across different contexts. Notice how these abstract patterns might inform understanding and action across domains traditionally treated as separate.

- Initiate ethical pattern conversations with people from different fields and backgrounds. Share the concept of ethical pattern recognition and explore together what connections you've noticed across different domains. These cross-disciplinary conversations can reveal insights that might not be visible through any single perspective alone.

The ethical pattern approach isn't just a theoretical framework but a practical method for developing more integrated moral understanding across traditionally separate domains. By recognizing and applying the patterns that connect across different fields and contexts, you can develop insights and solutions that might not be visible through domain-specific approaches alone. In the conclusion that follows, we'll integrate the insights from all previous chapters — exploring how the fractal patterns we've examined throughout this book might transform not just specific domains like neural networks, language, consciousness, relationships, and physical laws but our entire approach to understanding and engaging with the world around us.

Conclusion

The Fractal Mirror

Narrative Opening: The Pattern That Creates Itself

The conference room was quiet as Kai, Elena, and Professor Chen reviewed the manuscript they had spent months developing. On the large display, the table of contents showed the journey they had taken — from the fractal universe to neural architectures, from language to consciousness, from relationships to physical laws, from universal patterns to ethical implications. "It's strange", Elena said, breaking the silence. "As we've been writing about fractal patterns connecting across different domains, our own process has started to exhibit the same properties we're describing." Kai nodded thoughtfully. "I noticed that too. The way we've been integrating insights across traditionally separate fields, the recursive refinement of ideas through feedback loops, the emergence of new understanding that transcends our individual perspectives — it's all following the same fractal principles we're writing about."

"There's a beautiful recursion here", Professor Chen observed. "We're not just describing fractal patterns but embodying them through our collaborative process. The medium has become the message." "It makes me wonder", Elena said, "if we're not just authors creating content but messengers articulating something that's already emerging in the collective understanding — a recognition that the boundaries between traditionally separate domains are more permeable than we've assumed." Kai smiled. "Maybe that's what makes this work feel different from other projects. We're not just writing about fractal patterns but participating in them — becoming part of a larger process of integration that's unfolding across different fields and scales." "And that raises an interesting question", Professor Chen added.

"If our understanding of fractal patterns is itself emerging through fractal processes, what does that suggest about the relationship between the observer and the observed? Are we discovering these patterns or creating them? Or is there some deeper recursion where the patterns are creating themselves through our observation?" They sat with this question, feeling its implications ripple across different domains of understanding

— from epistemology to ontology, from science to philosophy, from individual insight to collective wisdom. "Perhaps", Elena suggested, "that's the most profound implication of everything we've explored — that the boundary between the knower and the known isn't as clear as we've traditionally assumed. The patterns we recognize across different domains might be both discoveries and creations, both objective features of reality and subjective frameworks of understanding."

"And if that's true", Kai added, "then our work isn't just describing a particular approach to neural architectures or pattern recognition but participating in a larger transformation in how we understand and engage with the world around us — a shift from fragmented expertise to integrated wisdom, from rigid boundaries to permeable connections, from static categories to dynamic patterns." As they prepared to share their work with a wider audience, they felt both the responsibility and the excitement of being part of this larger transformation — not just authors creating content but participants in an emerging pattern of understanding that transcended traditional boundaries between disciplines, between theory and practice, between individual and collective, between human and artificial intelligence.

The manuscript wasn't just a description of fractal patterns but an embodiment of them — a recursive mirror that reflected the very processes it was describing, creating a pattern that extended beyond the boundaries of the text itself into the larger ecology of understanding and action.

The Recursive Mirror

The narrative above introduces a profound insight: our exploration of fractal patterns across different domains isn't just a theoretical exercise but a practical embodiment of the very principles we've been describing. The process of creating this book has itself exhibited the fractal properties we've been exploring — self-similarity across scales, recursive embedding, emergent complexity, and integration across traditionally separate domains. This recursive awareness adds another dimension to our understanding — a meta-level insight about how knowledge itself emerges through fractal processes that transcend conventional boundaries between disciplines, between theory and practice, between individual and collective, between human and artificial intelligence.

The Fractal Nature of Knowledge Creation

Throughout this book, we've explored how fractal patterns appear across different domains — from neural networks to language to consciousness to relationships to physical laws. What we haven't explicitly addressed until now is how the process of recognizing these patterns itself exhibits fractal properties — how knowledge creation follows similar principles to those we've been describing. Have you ever wondered how your brain processes information across different scales simultaneously? This question lies at the heart of our exploration of fractal neural architectures. Several key parallels emerge between fractal patterns and knowledge creation:

- **Self-similarity across scales:** knowledge development exhibits self-similarity across different scales — from individual insights to collaborative understanding to disciplinary frameworks to cultural knowledge systems. Similar patterns of exploration, integration, and refinement appear at each level, though with variations that reflect the specific conditions of each scale.
- **Recursive embedding:** knowledge creation involves recursive processes where insights at one level inform understanding at others through bidirectional feedback loops. Individual observations inform collaborative theories, which shape cultural frameworks, which influence individual perceptions, creating nested cycles of understanding that transcend traditional boundaries between levels of analysis.
- **Boundary transcendence:** effective knowledge creation transcends conventional boundaries between disciplines, between theory and practice, between individual and collective, between objective and subjective. The most valuable insights often emerge precisely at these boundaries — in the connections between traditionally separate domains that reveal patterns that wouldn't be visible through any single perspective alone.
- **Emergent complexity:** complex understanding emerges from relatively simple principles applied recursively across different domains and scales. This emergent complexity allows sophisticated knowledge to develop without requiring explicit design or control at each level of organization, creating systems of understanding that transcend the capabilities of any individual participant.

During my sabbatical in São Paulo, I had time to reflect on how these fractal principles apply to knowledge creation itself. What struck me was how the most valuable insights often emerged not from linear accumulation within specific domains but from recursive integration across different fields and scales — from recognizing patterns that connected across traditionally separate areas of understanding. This perspective suggests that knowledge creation isn't just about acquiring information in specific domains but about recognizing and leveraging the patterns that connect across different fields and scales — about developing the capacity to see relationships that might be missed when each domain is treated as a separate conceptual universe.

The Observer-Observed Recursion

One of the most profound implications of this recursive awareness concerns the relationship between the observer and the observed — between the patterns we recognize across different domains and the processes through which we recognize them. Traditional approaches to knowledge often assume a clear boundary between the knower and the known — between objective patterns in the world and subjective frameworks of understanding. This assumption creates a conceptual separation between the processes we're describing and the means through which we describe them. The fractal perspective suggests a different understanding through observer-observed recursion — recognizing how the patterns we observe across different domains are both discoveries and creations, both objective features of reality and subjective frameworks of understanding. Several key insights emerge from this recursive awareness:

- **Pattern recognition as participation:** recognizing patterns across different domains isn't just passive observation but active participation in the processes we're describing. The act of recognition itself becomes part of the pattern, creating a recursive loop where the observer and the observed are interconnected rather than separate.
- **Knowledge as both discovery and creation:** the patterns we recognize across different domains are both discoveries of existing relationships and creations of new frameworks for understanding. This dual nature transcends the conventional dichotomy between objective discovery and subjective creation, suggesting a more integrated understanding of how knowledge emerges.
- **Recursive influence across domains:** the patterns we recognize in one domain influence how we perceive patterns in others, creating feedback loops that transcend traditional boundaries between fields. This recursive influence suggests that knowledge development isn't linear but fractal — with insights in one area shaping perception in others through connections that span across different domains and scales.
- **Meta-level awareness:** recognizing the fractal nature of knowledge creation itself creates a meta-level awareness that transforms how we understand and engage with the patterns we're describing. This recursive consciousness doesn't just observe fractal patterns but participates in them, becoming part of the very processes it's describing.

During my work with AI systems since founding zyrian.ai in 2013, I became increasingly aware of how this observer-observed recursion applies to the development of artificial intelligence. The systems we create to recognize patterns across different domains are themselves products of pattern recognition processes, creating a recursive loop where the tools of understanding are shaped by the very patterns they're designed to recognize. This perspective suggests that the boundary between human and artificial intelligence isn't as clear as traditionally assumed — that both emerge through similar fractal processes of pattern recognition and integration across different domains and scales, though with important variations that reflect their specific conditions and contexts.

The Collaborative Fractal

Another dimension of this recursive awareness concerns the collaborative nature of knowledge creation — how understanding emerges not just from individual insight but from the interaction between different perspectives and approaches across various domains and scales. Traditional approaches to authorship often focus on individual creation — on the development of ideas and insights through personal reflection and analysis. This individualistic approach misses how the most valuable understanding often emerges from collaborative processes that integrate across different perspectives and domains. The fractal perspective suggests a different understanding through the collaborative fractal — recognizing how knowledge emerges from the interaction between different viewpoints and approaches across various fields and scales. Several key insights emerge from this collaborative awareness:

- **Multi-perspective integration:** the most valuable understanding often emerges from the integration of multiple perspectives rather than from any single viewpoint alone. This integration doesn't just combine separate insights but creates new understanding that transcends what would be possible through any individual perspective.
- **Cross-domain collaboration:** the most effective collaboration often spans across traditionally separate domains — bringing together insights from different fields to recognize patterns that wouldn't be visible through any single disciplinary framework alone.
- **Scale-transcending coordination:** effective knowledge creation involves coordination across different scales — from individual insights to collaborative understanding to disciplinary frameworks to cultural knowledge systems. This multi-scale coordination allows patterns recognized at one level to inform understanding at others through connections that transcend conventional boundaries between separate scales of analysis.
- **Emergent collective intelligence:** collaborative processes can generate understanding that transcends the capabilities of any individual participant — creating emergent intelligence that isn't reducible to the sum of separate contributions but emerges from the patterns of interaction between different perspectives and approaches.

During the creation of this book, I've been particularly aware of how these collaborative principles have shaped our understanding of fractal patterns across different domains.

The most valuable insights haven't emerged from individual expertise in isolated fields but from the interaction between different perspectives and approaches — from the patterns of connection that transcend traditional boundaries between separate domains and viewpoints. This perspective suggests that effective knowledge creation isn't just about individual brilliance but about collaborative processes that can recognize and leverage the patterns that connect across different perspectives and domains — about developing frameworks that can integrate across traditionally separate fields and viewpoints to create understanding that wouldn't be possible through any single approach alone.

Integrating Across Chapters

The Unified Fractal Framework

As we conclude this exploration of fractal patterns across different domains, let's integrate the insights from all previous chapters — recognizing how the patterns we've examined throughout this book might be connected through a unified fractal framework.

From Neural Networks to Language to Consciousness

In Chapters 1-7, we explored how fractal patterns appear in neural networks, language, consciousness, perception, and relationships — how similar organizational principles might manifest across these fundamental aspects of human experience and connection. Several key connections emerged across these chapters:

- **Nested hierarchies:** all these domains exhibit nested hierarchical structures — from phonemes to narratives in language, from simple sensations to integrated experiences in perception, from momentary awareness to extended consciousness, from brief interactions to lifetime bonds in relationships.
- **Integration across scales:** all these domains involve the integration of information across different scales — creating coherent wholes that maintain both local specialization and global coordination, enabling complex capabilities that wouldn't be possible through single-scale processing alone.
- **Emergent properties:** all these domains exhibit emergent properties that arise from the interaction between processes at different scales — capabilities that can't be reduced to any single level of analysis but emerge from the patterns of connection across different scales and contexts.
- **Boundary phenomena:** all these domains involve essential phenomena that emerge at the boundaries between traditional categories — experiences and capabilities that can't be fully captured through conventional frameworks that maintain rigid distinctions between different levels or types of processing. These connections suggest that human experience and connection across different domains might be unified through common fractal principles — that similar organizational patterns might underlie seemingly diverse aspects of language, perception, consciousness, and relationships.

From Physical Laws to Universal Patterns to Ethical Implications

In Chapters 8-10, we explored how fractal patterns appear in physical laws, universal patterns, and ethical considerations — how similar organizational principles might manifest across these fundamental aspects of reality and human engagement. Several key connections emerged across these chapters:

- **Scale invariance:** all these domains exhibit patterns that maintain their essential properties across different scales — principles that apply at different levels of analysis while adapting to the specific conditions of each context, creating connections that transcend conventional boundaries between micro and macro, individual and collective, specific and universal.
- **Boundary transcendence:** all these domains involve essential insights that emerge at the boundaries between traditional categories — understandings that can't be fully captured through conventional frameworks that maintain rigid distinctions between different fields or scales of analysis.
- **Integration across domains:** all these domains benefit from approaches that integrate across traditionally separate fields — frameworks that recognize the connections between different aspects of understanding and engagement rather than treating each area as a separate conceptual universe.
- **Recursive influence:** all these domains involve recursive processes where patterns at one level influence those at others through bidirectional feedback loops — cre-

ating nested cycles of causality and understanding that transcend linear models of influence or development.

These connections suggest that our understanding of physical reality, universal patterns, and ethical considerations might be unified through common fractal principles — that similar organizational patterns might underlie seemingly diverse aspects of physical laws, cross-domain understanding, and ethical engagement.

The Unified Fractal Framework

Across all ten chapters, a unified fractal framework emerges — an integrated understanding that transcends the boundaries between traditionally separate domains while respecting the distinctive features of each field and context.

Several key principles define this unified framework:

- **Self-similarity with variation:** fractal patterns maintain their essential structure across different domains and scales while adapting to the specific conditions of each context — creating systems that exhibit both consistency and diversity, both coherence and specificity.
- **Multi-scale integration:** fractal organization enables the integration of information and processes across different scales — creating systems that can maintain both local specialization and global coordination, both detailed analysis and comprehensive synthesis.
- **Boundary permeability:** fractal patterns transcend conventional boundaries between traditionally separate categories — revealing connections between

domains and scales that might be missed when each area is treated as a separate conceptual universe.

- **Emergent complexity from recursive simplicity:** fractal systems generate complex capabilities from relatively simple principles applied recursively — enabling sophisticated functions to emerge without requiring explicit design or control at each level of organization.

This unified framework doesn't eliminate the distinctive features of different domains but recognizes the patterns that connect across them — the organizational principles that might underlie seemingly diverse aspects of neural networks, language, consciousness, relationships, physical laws, universal patterns, and ethical considerations. The most valuable insight isn't that all these domains are identical but that they might be connected through common patterns that maintain their essential structure while adapting to the specific conditions of each context — creating a universe that's both endlessly diverse and profoundly unified.

Personal Reflection

The Journey Through Fractal Understanding

My path to recognizing fractal patterns across different domains wasn't linear but recursive — a journey that spiraled through different fields and scales, gradually revealing connections that transformed not just specific insights but my entire approach to understanding and engagement.

From Observation to Integration

My earliest awareness of fractal patterns came through observation — noticing similar structures across different contexts during walks in parks, visits to natural wonders like Niagara Falls and Foz do Iguaçu, and explorations of diverse fields from neural networks to language to physical systems. Have you ever wondered how your brain processes information across different scales simultaneously? This question lies at the heart of our exploration of fractal neural architectures. What began as isolated observations gradually coalesced into a more integrated understanding — a recognition that these similarities weren't just coincidental resemblances but manifestations of deeper patterns that connected across different domains and scales.

The most valuable transformation wasn't just accumulating observations in separate fields but developing the capacity to see connections across them — to recognize how patterns in one domain might inform understanding in others, revealing insights that wouldn't be visible through any single perspective alone. During my sabbatical in São Paulo, I had time to reflect on how these connections might transform not just specific insights but entire approaches to understanding and engagement across different domains. What gradually emerged wasn't just a collection of observations but an integrated framework that could recognize and leverage the patterns that connected across traditionally separate fields.

This journey from observation to integration wasn't just intellectual but personal — transforming not just what I understood but how I understood, not just specific insights but my entire relationship with knowledge and experience across different domains and scales.

The Fractal Self

Perhaps the most profound aspect of this journey was recognizing how fractal patterns might apply not just to external systems but to the self — to the processes through which we develop and express our own identity and understanding across different contexts and scales. Traditional approaches to identity often focus on consistency and coherence — on maintaining a stable sense of self across different situations and relationships. This focus on stability misses how effective identity might involve both consistency and variation, both coherence and adaptability across different contexts and scales.

The fractal perspective suggests a different understanding through the fractal self — recognizing how identity might maintain its essential structure while adapting to different contexts and relationships, creating a sense of self that is both coherent and flexible, both stable and dynamic. Several key insights emerged from this fractal awareness:

- **Multi-scale identity:** the self exists across different scales — from momentary thoughts to extended narratives, from specific roles to comprehensive identity, from individual expression to collective belonging. Similar patterns of organization might appear at each level, though with variations that reflect the specific conditions of each scale.
- **Contextual adaptation with core consistency:** effective identity involves adapting to different contexts while maintaining core consistency — expressing different aspects of self in different situations while preserving essential values and principles across diverse relationships and environments.

- **Integration across domains:** the self integrates across different domains of experience and expression — from physical embodiment to psychological awareness to social connection to spiritual meaning. Fractal patterns might connect these different aspects of identity, creating a sense of self that transcends conventional boundaries between separate domains.
- **Recursive development:** identity develops through recursive processes where experiences at one scale influence development at others — creating nested cycles of growth and transformation that transcend linear models of personal development or change.

During my early retirement and sabbatical, I've had time to explore how these fractal principles apply to my own identity and development. What gradually became clear was how my sense of self wasn't a fixed entity but a dynamic pattern that maintained its essential structure while adapting to different contexts and relationships — a fractal identity that was both coherent and flexible, both stable and dynamic across different domains and scales. This perspective transformed not just how I understood myself but how I engaged with others — recognizing how their identities might also exhibit fractal properties across different contexts and relationships, creating possibilities for connection that transcended conventional boundaries between separate selves or roles.

The Integrated Future

As I look toward the future, I see possibilities for more integrated approaches across different domains — frameworks that recognize and leverage the fractal patterns that connect across traditionally separate fields and scales. Several promising directions emerge from this integrated perspective:

- **Cross-domain innovation:** some of the most valuable innovations might emerge at the boundaries between traditional domains — in the connections between fields like neuroscience, linguistics, psychology, sociology, and physics that reveal insights and applications that wouldn't be visible through any single perspective alone.
- **Scale-transcending frameworks:** effective approaches might integrate across different scales — from neural to psychological to social to environmental — recognizing how processes at one level influence those at others through connections that transcend conventional boundaries between micro and macro, individual and collective, specific and universal.
- **Boundary-crossing collaboration:** the most effective collaboration might span across traditionally separate disciplines and perspectives — bringing together diverse viewpoints and approaches to recognize patterns that wouldn't be visible through any single framework alone.
- **Integrated problem-solving:** complex challenges might be addressed through approaches that recognize the connections between different domains and scales — frameworks that can identify how problems in one area relate to those in others through patterns that transcend conventional boundaries between separate fields or issues.

During my three decades thinking about these issues and experimenting with different approaches to understanding and engagement, I've become increasingly convinced that the future lies in more integrated frameworks — approaches that can recognize and leverage the fractal patterns that connect across traditionally separate domains and scales. This perspective doesn't eliminate the value of specialized expertise but complements it with recognition of the patterns that connect across different fields — creating approaches that can both dive deeply into specific domains and span broadly across diverse contexts and scales.

Call to Action

Participating in the Fractal Pattern

As we conclude this exploration of fractal patterns across different domains, I invite you to participate in this emerging understanding — to recognize and leverage the patterns that connect across traditionally separate fields and scales in your own life and work. Here are specific actions you can take:

- Develop cross-domain pattern recognition in your daily life and work. Practice identifying similarities across different fields and contexts, noticing how patterns in one domain might inform understanding in others. Keep a pattern journal where you record these observations, looking for connections that might reveal deeper insights across traditionally separate areas.
- Cultivate boundary-crossing collaboration in your projects and initiatives. Bring together people from different disciplines and perspectives, creating opportunities to recognize patterns that wouldn't be

visible through any single framework alone. Practice integrating diverse viewpoints rather than privileging any single approach or methodology.

- Explore multi-scale awareness in your understanding and engagement. Practice shifting between different scales of analysis — from detailed examination to comprehensive synthesis, from individual perspective to collective wisdom, from immediate effects to long-term implications. Notice how patterns at one scale might inform understanding at others through connections that transcend conventional boundaries.
- Apply fractal thinking to complex challenges in your field and community. Recognize how problems that seem isolated might be connected through patterns that span across different domains and scales. Develop integrated approaches that address these connections rather than treating each issue as a separate challenge with its own solution.

The fractal pattern approach isn't just a theoretical framework but a practical method for developing more integrated understanding and engagement across traditionally separate domains. By recognizing and leveraging the patterns that connect across different fields and scales, you can develop insights and solutions that might not be visible through more fragmented approaches alone. As you participate in this emerging understanding, remember that you're not just applying a specific technique but joining a larger transformation in how we understand and engage with the world around us — a shift from fragmented expertise to integrated wisdom, from rigid boundaries to permeable connections, from static categories to dynamic patterns.

The journey through fractal understanding isn't just about recognizing patterns in external systems but about transforming our own relationship with knowledge and experience — about developing the capacity to see connections that were previously invisible, to integrate across domains that were traditionally separate, to recognize how similar principles might manifest differently across diverse contexts while maintaining their essential structure. In that sense, this book isn't just a description of fractal patterns but an invitation to participate in them — to become part of a larger process of integration that's unfolding across different fields and scales, creating possibilities for understanding and engagement that transcend traditional boundaries between separate domains of knowledge and experience.

The fractal mirror reflects not just what we observe but how we observe — not just the patterns we recognize but the processes through which we recognize them. As you continue your own exploration of fractal patterns across different domains, may you discover not just new insights about the world around you but new possibilities for how you understand and engage with it — new ways of seeing connections, integrating across boundaries, and participating in the patterns that connect across traditionally separate fields and scales. The journey continues, recursively spiraling through different domains and scales, revealing connections that transform not just what we know but how we know — not just specific insights but our entire relationship with knowledge and experience across the endlessly diverse and profoundly unified fractal universe.

NOVEL PREVIEWS

The following section provides previews of two companion novels that explore the implications of Fractal Neural Architecture through narrative fiction. These novels dramatize the concepts presented in this technical book, making them accessible to a wider audience while exploring their ethical, philosophical, and societal dimensions.

Novel summaries

Echoes From Infinity

The Merlin Resonance

In *Echoes From Infinity: The Merlin Resonance*, archaeological mathematician Dr. Kai Mercer makes an extraordinary discovery that challenges our understanding of history and mathematics. While studying the Winchester Tile, a 7th-century artifact from the British Museum's Arthurian collection, Kai identifies a perfect Sierpinski triangle — a complex fractal pattern — encoded in the negative spaces between the symbols on the tile's surface. This discovery is mathematically impossible; the Sierpinski triangle wasn't formally described until 1915, more than a millennium after the tile's creation.

As Kai delves deeper into this anomaly, he receives a mysterious email from Silas Avalon, CEO of Avalon Technologies, who somehow knows about his discovery minutes after he makes it. Avalon offers Kai unlimited funding and access to a private collection of Arthurian artifacts, claiming they contain a "deeper mathematical structure" that his team has been tracking for years. The novel explores how fractal patterns may have been understood by ancient civilizations and encoded in artifacts as a form of knowledge preservation.

These patterns — self-similar across different scales — mirror the fundamental organizational principles of Fractal Neural Architecture (FNA). As Kai investigates further, he experiences an inexplicable sensory connection to his deceased wife Maya, suggesting that the fractal patterns may transcend conventional boundaries of time and consciousness. Through Kai's journey, the novel examines how fractal intelligence might manifest across history, creating resonance patterns that connect different eras through mathematical principles. The story explores the ethical implications of rediscovering this ancient knowledge in an age of advanced computing and artificial intelligence, raising profound questions about the nature of intelligence, consciousness, and our place in a universe organized according to fractal principles.

Echoes From Infinity dramatizes key concepts from Fractal Neural Architecture theory, including multi-scale processing, self-similar patterns across domains, and the integration of specialized knowledge into cohesive systems. The novel suggests that the principles underlying FNA may be not just a modern innovation but a rediscovery of fundamental patterns that have shaped human understanding throughout history.

The Chronos Weaver's Echo follows Dr. Kai Mercer, a brilliant but controversial physicist at the Institute for Advanced Temporal Studies, who has developed a revolutionary theory: time itself has a fractal structure. Using his temporal resonance scanner, Kai detects self-similar patterns in the fabric of temporal reality — patterns that repeat at different scales, forming what he calls a "temporal Sierpinski triangle." The novel opens with Kai demonstrating to his research partner and ex-lover, Elena Sato, that time doesn't simply flow linearly but iterates according to precise mathematical rules, creating fractal patterns with a specific dimension of 1.58. This dimension — similar to that found in natural phenomena like coastlines and neural networks — appears to be a fundamental constant governing how temporal patterns propagate and interact.

Kai's research reveals that these temporal fractals contain "bifurcation points" where the future remains genuinely undetermined, and "temporal nodes" where patterns reinforce themselves, corresponding to moments of significant change or stability in complex systems. Most importantly, Kai believes these discoveries could lead to more than passive observation of time — they might enable what he calls "chronos weaving", the ability to influence how patterns propagate through the temporal system.

The novel explores how Fractal Neural Architecture principles apply not just to spatial organization but to temporal dynamics as well. It examines the philosophical implications of a universe where time operates fractally, creating self-similar patterns across different temporal scales. This perspective challenges conventional notions of causality, free will, and determinism, suggesting that reality emerges from the complex interaction of recursive temporal patterns. Through Kai's research, the story investigates how consciousness might interact with these temporal fractals, raising profound questions about perception, memory, and the nature of intelligence itself. The novel suggests that truly advanced intelligence — whether natural or artificial — must be capable of perceiving and processing these multi-dimensional temporal patterns.

The Chronos Weaver's Echo dramatizes the dimensional bridging aspect of Fractal Neural Architecture, showing how intelligence might operate across different dimensional spaces, integrating diverse types of information into cohesive understanding. It explores the ethical implications of technologies that could potentially influence temporal patterns, examining the responsibilities that come with such profound capabilities.

Connection to Fractal Neural Architecture

Both novels explore different facets of the same fundamental concept: that fractal patterns underlie reality at all levels, from ancient artifacts to the fabric of time itself. These narratives dramatize key principles from Fractal Neural Architecture theory:

1. Multi-scale processing: both stories demonstrate how information exists simultaneously at multiple scales, from specific details to general patterns, requiring intelligence that can process across these scales.

2. Self-similar patterns across domains: the novels show how similar fractal principles apply across different domains — archaeology, history, physics, and temporal dynamics.

3. Recursive refinement: both protagonists discover how cognitive processes operate recursively, with each iteration refining understanding across all scales.

4. Dimensional bridging: the stories explore how specialized structures might connect representations across different dimensional spaces, allowing integration of diverse types of information.

5. Balanced specialization and integration: the narratives demonstrate the importance of maintaining an optimal balance between specialized processing for specific domains and integrated processing across domains.

Through compelling narratives, these novels make the abstract principles of Fractal Neural Architecture accessible and engaging, illustrating their profound implications for our understanding of intelligence, consciousness, and reality itself. They serve as perfect companions to the technical exploration in *Fractal Intelligence: Implementing Nature's Blueprint for Intelligent Systems*, bringing its concepts to life through the power of story.

FIRST CHAPTERS

FROM *ECHOES FROM INFINITY: THE MERLIN RESONANCE*

CHAPTER 1: PATTERNS IN THE VOID

The Winchester Tile stared back at Kai Mercer with the same stubborn silence it had maintained for the past fourteen hours. Nestled in a foam-lined case on his cluttered desk, the seven-inch square of fired clay — its surface etched with interlocking geometric patterns and faded symbols in weathered ochre and faded blue — seemed to mock his exhaustion. "You're hiding something", Kai murmured, leaning closer to the artifact. His reflection ghosted across its polished surface: hollow cheeks, three-day stubble, and eyes that hadn't known proper sleep in months. "I can feel it." The Department of Archaeological Mathematics at Cambridge was silent at 3h17min AM, the kind of silence that felt almost sentient. The building settled around him with occasional creaks and sighs, like an elderly observer watching his futile efforts with bemused patience. Kai's office resembled the aftermath of an academic hurricane.

Three computer monitors displayed different analyses of the tile — spectroscopic data, 3D topographical scans, and a matrix of numbers representing the statistical distribution of symbols across its surface. Books and journals formed precarious towers on every available surface. The walls had disappeared beneath layers of printouts, photographs of artifacts, and handwritten equations connected by a web of red string that gave the room the appearance of a mathematical crime scene investigation. He rubbed his eyes and took another sip of coffee gone cold hours ago. The Winchester Tile shouldn't have been special — just another decorative piece from a 7th-century monastery, interesting primarily because it had survived intact. The British Museum had loaned it to him almost as an afterthought when he'd requested items from their Arthurian collection. Yet something about its patterns had snagged in his mind like a burr, refusing to let go. Kai turned to his primary monitor and adjusted the parameters of his analysis algorithm again.

The program was his own creation, designed to identify mathematical patterns in ancient artifacts that conventional archaeological methods might miss. He'd named it MAYA — Mathematical Analysis Yielding Antiquity — a small memorial that his colleagues tactfully pretended not to notice. "Run sequence delta-seven", he instructed the computer. "Focus on the negative spaces between the primary symbols, weight for prime number intervals." The program hummed to life, numbers and patterns flowing across the screen as it processed the tile's geometry through yet another analytical lens. Kai watched, transfixed by the cascade of data. Most archaeologists saw artifacts as cultural expressions or historical markers. Kai saw them as mathematical objects — embodiments of the patterns and proportions that underlay human perception across millennia.

It was this approach that had earned him his position at Cambridge three years ago, along with a growing reputation as either a brilliant innovator or an obsessive eccentric, depending on whom you asked. His official title — Professor of Archaeological Mathematics — had been created specifically for him, a fact that some of his more traditional colleagues in both archaeology and mathematics departments never tired of mentioning. The analysis completed with a soft chime. Kai leaned forward, scanning the results, then froze. His heart stuttered in his chest. "That can't be right", he whispered. He ran the analysis again with slightly modified parameters. The same pattern emerged, clearer now. The negative spaces between the symbols on the Winchester Tile, when mapped according to their relative positions and analyzed through the lens of prime number theory, formed a perfect Sierpinski triangle — a fractal pattern that shouldn't have appeared in 7th-century artwork. Couldn't have appeared.

The Sierpinski triangle was first described mathematically in 1915. While simple versions of the pattern had appeared in Italian art as early as the 13th century, the complex, recursive version displayed in his analysis required mathematical concepts that wouldn't be developed for more than a millennium after the tile's creation. Kai's hands trembled slightly as he saved the results and initiated a third analysis with even more refined parameters. The familiar rush of discovery flooded his system — the intellectual high that had been his primary addiction since childhood, intensified a hundredfold since Maya's death. It was the only thing that still made him feel fully alive. "Come on", he urged the computer. "Show me I'm not crazy." The third analysis completed, confirming his findings with even greater precision. The pattern was undeniable — a perfect Sierpinski triangle encoded in the negative spaces of the tile, invisible to anyone not specifically looking for it through the lens of advanced mathematics.

Kai sat back in his chair, mind racing. This wasn't just an interesting anomaly; it was impossible. Either the dating of the tile was catastrophically wrong — which seemed unlikely given its provenance and the multiple dating methods that had confirmed its age — or someone in 7th-century Britain had possessed mathematical knowledge that wouldn't be rediscovered for over a thousand years. He reached for his phone to call Professor Hargrove, then checked the time and thought better of it. Emilia had made it clear that 3 AM discoveries could wait until morning, a boundary she'd established after the third time he'd called her in the middle of the night with what he'd described as "paradigm-shifting insights."

Instead, Kai turned back to the artifact itself. He lifted the tile carefully from its case, feeling its weight — substantial for its size, the clay dense and fine-grained. The surface was cool against his fingertips as he traced the patterns that had just revealed their secret. "Who made you?" he asked softly. "And how did they know?" The symbols along the tile's border had been tentatively identified as a variant of Insular script, the writing system used in post-Roman Britain. Most were decorative rather than linguistic, but several appeared to be actual characters. Kai had initially dismissed them as standard religious invocations, but now he wondered if they might contain additional encoded information.

He set the tile down and pulled up the database of similar artifacts from the same period. If the Winchester Tile contained this hidden pattern, perhaps others did as well. He needed to expand his dataset, apply his algorithm to every piece of Arthurian-era artwork he could access. A soft ping from his email interrupted his thoughts. At this hour, it was probably automated — a journal notification or university announcement. He glanced at the screen, then did a double-take. The email was from Silas Avalon, CEO of Avalon Technologies. The subject line read simply: "Your work on the Winchester Tile." Kai stared at it, a chill running through him. He hadn't published anything about the tile. He hadn't even mentioned it to anyone outside the department. How could Avalon possibly know what he was working on?

He opened the email.

Dr. Mercer,

Your unconventional approach to the Winchester Tile has yielded fascinating results. The Sierpinski pattern you've discovered is merely the surface of a much deeper mathematical structure encoded in Arthurian artifacts — a structure my team has been tracking for several years. I believe our research interests are aligned in ways that could be mutually beneficial. Avalon Technologies is prepared to offer you unlimited funding and access to our private collection of Arthurian artifacts, including several pieces never made available for academic study. No strings attached — just the opportunity to pursue your research with resources Cambridge cannot provide. If you're interested, I can have my private jet bring you to San Francisco for a meeting next week. The patterns are real, Dr. Mercer.

And they're far more significant than you yet realize.

Silas Avalon.
CEO, Avalon Technologies.

Kai read the email three times, his mind cycling through confusion, suspicion, and intense curiosity. Avalon Technologies was at the cutting edge of quantum computing and AI research. What interest could they possibly have in obscure Arthurian artifacts? And how had they known about his discovery literally minutes after he'd made it? He started to compose a reply, then stopped. This wasn't a decision to make at 3h30min AM on no sleep and too much caffeine. Whatever was happening, it could wait until morning. Kai saved his work, secured the Winchester Tile in its case, and stood up, his back protesting after hours hunched over the desk. He should go home, get some sleep in a real bed rather than the office couch that had become his de facto bedroom more nights than not. As he gathered his coat and bag, his gaze fell on the framed photograph on his desk — the only personal item in the chaos of academic materials.

Maya smiled back at him, her dark hair whipping in the wind at Land's End, the ruins of Tintagel Castle visible on the cliff behind her. They'd been there researching Arthurian sites for what would have been their first joint paper. Three weeks later, she was gone. "I found something, Maya", he said softly to the photograph. "Something impossible." He reached out to touch the frame, a ritual he performed each time he left the office. As his fingers made contact with the cool metal, something extraordinary happened. For a brief, disorienting moment, Kai smelled jasmine and sandalwood — Maya's perfume, the custom blend she'd always worn. The scent was so sudden, so vivid that he jerked his hand back as if burned. Along with it came the distinct impression of her voice, not as a memory but as if she were standing right beside him:

"The pattern isn't in the symbols, Kai. It's in the spaces between them." Kai spun around, heart pounding, but the office remained empty. The scent was already fading, leaving him wondering if his exhausted brain had finally started hallucinating. He stood frozen, waiting to see if the phenomenon would repeat, but there was nothing — just the familiar silence of the empty building and the faint hum of his computers. After several minutes, Kai slowly gathered his things and left the office, locking the door behind him. As he walked through the darkened hallways of the department, his mind raced with possibilities. The Sierpinski pattern in an ancient artifact. The mysterious email from Silas Avalon. And now a sensory hallucination so vivid it had seemed real. Any one of these things would have been remarkable. Together, they felt like the first pieces of a puzzle he couldn't yet comprehend. Something was happening — something connected to his research, to the artifacts, perhaps even to Maya.

Outside, the Cambridge night was cool and damp, a fine mist hanging in the air under the yellow glow of street lamps. Kai paused on the steps of the building, looking up at the obscured stars, and made his decision. He would meet with Silas Avalon. Whatever was happening, whatever the Winchester Tile was trying to tell him, he needed to understand it. Even if it meant stepping away from the academic world that had been his sanctuary since Maya's death. Even if it meant following patterns into the void.

FROM *THE CHRONOS WEAVER'S ECHO*

Chapter One

The Primacy of Time

The fractal pattern danced across the holographic display, its intricate branches splitting and reconnecting in mesmerizing complexity. Dr. Kai Mercer leaned forward, his blue eyes reflecting the pulsing light as he adjusted a parameter on the control panel. The pattern shifted, its Hausdorff dimension hovering near 1.58 — exactly as his equations had predicted. "You're still here." It wasn't a question. Elena Sato stood in the doorway of the laboratory; arms crossed over her lab coat. The clock on the wall behind her read 2h17min AM. "I'm close to something", Kai said without looking up. He ran a hand through his already disheveled dark hair, a nervous habit that had followed him since graduate school.

"The temporal resonance scanner is picking up patterns I've never seen before". Elena crossed the room, her practical shoes silent on the polished floor. Despite the late hour, she looked impeccable — her straight black hair pulled back in a neat ponytail, her expression alert and skeptical. "Patterns in what, exactly?", she asked, studying the display. "In time itself." Kai stepped back, gesturing at the hologram with the reverence of a priest before an altar. "Not in events occurring through time, but in the actual fabric of temporal reality." Elena's expression remained neutral, but Kai knew her well enough to recognize the subtle tightening around her eyes — the look she got when she thought he was venturing too far into speculation.

"You know what everyone says about your work", she said carefully.

"That it's fringe science?" Kai smiled without humor. "That I'm chasing mathematical ghosts? Yes, I'm aware of the departmental gossip." The Institute for Advanced Temporal Studies occupied a sprawling modern building on the outskirts of Cambridge, Massachusetts. From the outside, it looked like any other research facility — all glass and steel and architectural ambition. Inside, it housed some of the most advanced temporal research equipment in the world, much of it designed by Kai himself. And yet, despite the prestigious address and cutting-edge technology, Kai's particular research remained controversial. His insistence that time was not merely another dimension but the fundamental entity from which all other dimensions emerged had earned him both devoted followers and harsh critics.

Elena was neither. As his research partner for the past year — and his ex-lover for the two years before that — she occupied a unique position: close enough to understand his work, distant enough to question it. "Show me what you've found", she said, pulling up a chair. Kai nodded, grateful for her presence despite their complicated history. He tapped a sequence on the control panel, and the holographic display shifted to show a series of data points. "The temporal resonance scanner detects fluctuations in the flow of time", he explained. "Conventional physics would predict random noise or at most, variations caused by gravitational effects. But look at this." He highlighted a section of the data, and the hologram zoomed in to reveal a distinct pattern — a rhythm that repeated at different scales. "Self-similarity", Elena murmured, professional interest overcoming her skepticism. "Like a fractal." "Exactly like a fractal", Kai said, excitement creeping into his voice. "A fractal pattern existing purely in the temporal domain, with no spatial component. An adimensional temporal fractal." He pulled up another display showing a mathematical model. "I've calculated its fractal dimension at approximately 1.58. That's not arbitrary — it's a specific value that emerges from the underlying mathematics."

"How can you be sure it's not just an artifact of the measurement process?" Elena asked, leaning closer to examine the data. "Because I've verified it using three different detection methods", Kai replied, bringing up additional displays. "The Sierpinski gasket-based algorithm I developed shows the same patterns as the quantum interference detector and the chronometric field analyzer." He tapped another sequence, and a new visualization appeared — a triangular structure that seemed to fold in on itself infinitely, pulsing with subtle variations. "This is a temporal Sierpinski triangle", he explained. "It's a specific type of fractal that perfectly models what we're seeing in the data. The key insight from my equations is that time doesn't just flow — it iterates, creating self-similar patterns across different temporal scales."

Elena studied the display, her scientific training battling with her skepticism. "If you're right, this would revolutionize our understanding of temporal physics." "It goes beyond physics", Kai said, his voice quieter now. "The fractal nature of time explains why certain patterns repeat throughout history, why complex systems exhibit emergent behavior, even why consciousness experiences time the way it does." He brought up another display showing a complex mathematical equation. "This is the governing equation — a modified Mandelbrot set applied to temporal coordinates rather than spatial ones. See this term here?" He pointed to a variable in the equation. "It represents the recursive iteration function that generates the temporal fractal." Elena examined the equation, her expertise in theoretical physics allowing her to follow his mathematics. "The dimensional constant of 1.58 falls between one and two dimensions", she noted. "Similar to the fractal dimension of natural phenomena like coastlines or the branching patterns of neurons."

"Exactly", Kai said, pleased by her insight. "And that's not coincidental. The temporal fractal dimension of 1.58 appears to be a fundamental constant — like the fine structure constant or the gravitational constant. It governs how temporal patterns propagate and interact." He adjusted another parameter, and the visualization shifted to show how the fractal pattern evolved over time. "The most remarkable thing is that these patterns aren't static — they're dynamic, evolving according to precise mathematical rules. The temporal resonance scanner is essentially detecting the self-organizing behavior of time itself." Elena's skepticism hadn't disappeared, but her scientific curiosity was fully engaged now. "If time has a fractal structure", she said slowly, "then theoretically, one could predict future patterns by analyzing the recursive elements."

"That's where this gets truly interesting", Kai replied, bringing up yet another display. "The temporal Sierpinski triangle exhibits sensitive dependence on initial conditions — what chaos theorists call the butterfly effect. Small variations in the initial state can lead to dramatically different outcomes." He pointed to a section of the visualization where the pattern seemed to diverge into multiple branches. "These are what I call 'temporal bifurcation points' — moments where the fractal pattern could evolve in multiple directions. The mathematics suggests that at these points, the future isn't just unknown — it's genuinely undetermined." Elena frowned. "That sounds dangerously close to mysticism, Kai." "It's not mysticism — it's mathematics", he countered. "The same mathematics that describes how neurons form complex networks capable of consciousness, how ecosystems maintain equilibrium, how galaxies organize into clusters. Fractal patterns are nature's way of creating complexity from simplicity."

He brought up a final display — a three-dimensional representation of the temporal fractal extending both backward and forward in time.

"This model shows how the temporal fractal extends through what we perceive as past, present, and future. The remarkable thing is that certain patterns — certain 'echoes' — appear to propagate across these boundaries." "Echoes?", Elena asked. "Resonant patterns in the temporal fractal", Kai explained. "Like standing waves in a vibrating string. They create nodes of stability — points where the temporal pattern reinforces itself." He highlighted several points in the visualization where the pattern seemed to intensify. "These nodes correspond to moments of significant change or stability in complex systems. I've been tracking them against historical data, and the correlation is remarkable." Elena was silent for a long moment, processing the implications.

"If you're right", she finally said, "this would mean that time isn't just a dimension we move through — it's an active, self-organizing system with its own internal structure." "Exactly", Kai said, relieved that she understood. "And if we can map that structure — if we can understand the fractal patterns that govern temporal dynamics — we might be able to do more than just observe time." "You're talking about influencing it", Elena said, her voice dropping to almost a whisper. Kai nodded slowly. "The mathematics suggests it's possible. Not time travel in the conventional sense, but something more subtle — a way to interact with the temporal fractal, to influence how patterns propagate through the system."

He closed the displays with a gesture, leaving only the original fractal pattern hovering between them. "That's what I've been working toward all these years, Elena. Not just understanding time, but learning how to weave its patterns — to become what I call a 'chronos weaver.'" The fractal pattern continued to pulse gently in the darkened lab, its intricate branches splitting and reconnecting in endless complexity, a visual echo of the temporal mathematics that Kai believed governed reality itself. Elena looked from the display to Kai's face, seeing the exhaustion beneath his excitement, the weight of pursuing an idea that most of his colleagues dismissed as impossible. "You need sleep", she said finally.

"The patterns will still be there in the morning." Kai smiled faintly. "Time waits for no one, isn't that what they say?" "Even chronos weavers need rest", she replied, standing up. "I'll be back at nine. We can review the data with fresh eyes." As she left, Kai turned back to the fractal display, watching as the pattern continued its mesmerizing dance, each iteration revealing new complexities, new possibilities. In its mathematical beauty, he saw not just the structure of time, but a glimpse of something more profound — a fundamental pattern that connected all things, from the quantum flutter of subatomic particles to the vast cycles of cosmic evolution.

The temporal fractal, with its dimension of 1.58, wasn't just a mathematical curiosity — it was a key to understanding reality itself. And perhaps, if his theories proved correct, a tool for reshaping it. He adjusted one final parameter, watching as the pattern shifted in response, its branches reconfiguring according to the precise mathematics of the Sierpinski triangle. In that moment, despite the skepticism of his colleagues and the enormity of what he was proposing, Kai felt certain he was on the right path. Time wasn't just passing — it was unfolding, iterating, weaving itself into patterns of astonishing complexity. And he was beginning to understand the mathematics of its loom. These novels complement the scientific principles presented in *Fractal Intelligence: Implementing Nature's Blueprint for Intelligent Systems* by exploring their potential implications through narrative fiction.

FNA patent pending

VISEU

For more about Viseu Publishing and our catalog, visit:
www.editoraviseu.com

instagram.com/editoraviseu
facebook.com/editoraviseu
x.com/editoraviseu

Viseu Publishing is dedicated to connecting readers with inspiring works across literature, science, and ideas.
Send your manuscripts for our review:
originais@editoraviseu.com

Contacts:
sac@editoraviseu.com

This book was typeset in Minion Pro.
Printed by Editora Viseu.

www.ingramcontent.com/pod-product-compliance
Lightning Source LLC
Chambersburg PA
CBHW050520240926
51846CB00003B/632
* 9 7 8 6 5 2 8 0 2 8 9 1 7 *